The Best on Quality

THE BEST ON QUALITY

Edited by John D. Hromi

Book Series of
International Academy for Quality

Vol. 5

ASQC Quality Press
Milwaukee, Wisconsin

The Best on Quality, Volume 5
John D. Hromi, editor

© 1995 by ASQC
All rights reserved. No part of this book may be reproduced in any form or by any means, electronic, mechanical, photocopying, recording, or otherwise, without the prior written permission of the publisher.

10 9 8 7 6 5 4 3 2 1

ISSN 0936-160X
ISBN 0-87389-286-0

ASQC Mission: To facilitate continuous improvement and increase customer satisfaction by identifying, communicating, and promoting the use of quality principles, concepts, and technologies; and thereby be recognized throughout the world as the leading authority on, and champion for, quality.

For a free copy of the ASQC Quality Press Publications Catalog, including ASQC membership information, call 800-248-1946.

Printed in the United States of America

 Printed on acid-free recycled paper

ASQC
Quality Press
611 East Wisconsin Avenue
Milwaukee, Wisconsin 53202

Contents

Introduction: The IAQ National and International Quality Awards Project

Chapter 1:
The Deming Prize
BY YOSHI KONDO, HITOSHI HUME, AND SHOICHI SCHIMIZU

Chapter 2: American Quality Awards:
Profiles in Excellence
BY RAYMOND WACHNIAK

Chapter 3: The Malcolm Baldrige National Quality Award: Seven Years of Progress, 7000 Lessons Learned
BY A. BLANTON GODFREY

Chapter 4: Federal Quality Institute Awards for Federal Government Groups and Employees
BY WILLIAM A. J. GOLOMSKI

Chapter 5: NASA Excellence Award for Quality and Productivity
BY RAYMOND WACHNIAK

Chapter 6: IIE Award for Excellence in Productivity Improvement
BY KENNETH E. CASE

Chapter 7: RIT/*USA TODAY* Quality Cup Award for Individuals and Teams
BY WILLIAM A. J. GOLOMSKI.

Chapter 8: The European Quality Award
BY JOHN A. GOLDSMITH

Chapter 9: The British Quality Award Scheme
BY ROY KNOWLES

Chapter 10: The Scandinavian Quality Awards
BY KERSTIN JÖNSON AND ASBJØRN AUNE

Chapter 11: Prix Qualité France

Chapter 12: The Italian FONTI Quality Award for Small/Medium-Size Service Companies

Chapter 13: Developing a National User-Friendly Certification and Awards Scheme

Chapter 14: Preliminary Information on the Union for Czech and Slovak Quality Awards

Chapter 15: Report on the Argentine National Quality Award Program

Chapter 16: National Quality Awards: A Developing Country Perspective

Chapter 17: A Critical Review of Methods for Quality Awards and Self-Assessment

Section II: Other Papers Dealing with Quality Awards

Chapter 18: The Australian Quality Awards

Preface

It is the policy of the International Academy for Quality (IAQ) to encourage its members, known as *academicians*, to foster as individuals the development of quality control (QC) methods, quality systems, and their management. In doing this, each member can draw on the knowledge and insight of his or her peers to better serve the good cause.

In 1988, IAQ presented the first in a series of books titled *The Best on Quality.* The first and subsequent volumes contain results of project work, carried out by groups of members to shed light onto important quality-relevant problems of today and tomorrow. A group, headed by academician Tito Conti, chose as its project a study of national and international quality awards.

Companies throughout the world compete for quality awards. There are those who are singled out for special recognition because they excel above others in satisfying the award's criteria; but it has been demonstrated that even those who have not won a prize are better for having tried.

Perhaps the best known quality awards are the Deming Prize, the Malcolm Baldrige National Quality Award (MBNQA), and the European Quality Award. Nevertheless, the creation and implementation of this form of recognition remains a dynamic activity – new awards are being created, older awards are being revised continually, and others are still in the planning stages. Since the information for this volume was accumulated, some late developments include announcements of national quality awards by the Czech Republic and Singapore. The initiative for developing and implementing an award procedure varies; sponsorship may be governmental, civic (nongovernmental), professional, and private. Hence, this volume is an effort to share with the reader the evolution and – in some cases – the experiences arising out of actual award participation.

Section 1 presents those papers that were assembled by the academy's group whose project dealt with national and international quality awards. Section 2 offers additional papers on the same topic that were submitted to the editor. Section 3 deals with more localized awards. Appendix A provides a list of quality awards that are in place in each state. Appendix B provides a list that summarizes the status of foreign quality awards. Appendix C contains the IAQ Constitution, the academy's history, and its membership.

As this volume illustrates, quality awards schema encourages competition, it provides a basis for self-assessment, and offers guidelines to suppliers dedicated to meeting customer expectations.

John D. Hromi
Academician
Editor, *The Best on Quality, Volume 5*

About the Editor

John D. Hromi, D. Engr., is an internationally known educator, practitioner, and consultant in quality control. He is Professor Emeritus and founder of the John D. Hromi Center for Quality and Applied Statistics, Rochester Institute of Technology. An ASQC Fellow, he is a former ASQC president and board chairman. He has received ASQC's Automotive Division's C. C. Craig Award for excellence in quality control and reliability, E. L. Grant Award for creating education programs that enhance quality assurance ethics and practices, and the Edwards Metal for his work in modern quality management. He is a judge for the RIT/*USA TODAY* Quality Cup Award and for the New York State Excelsior Award.

Acknowledgments

The editor wishes to express his appreciation to the following members of IAQ who served in the assembling and selection of articles for this book.

T. CONTI – PROJECT LEADER
K. JONSON – COORDINATOR, SCANDINAVIA
Y. KONDO – COORDINATOR, JAPAN
K. S. STEPHENS – COORDINATOR, DEVELOPING COUNTRIES
R. WACHNIAK – COORDINATOR, U.S.A.

He also thanks all the authors and publishers who offered contributions or gave rights for publishing. Not to be forgotten are the data entry clerks at the Rochester Institute of Technology (RIT). Their valuable assistance is appreciated.

JOHN D. HROMI

Introduction: The IAQ National and International Quality Awards Project

In the last few years, two issues have dominated the stage in the quality field. The first, a theme of particular interest in Europe, is certification with reference to the ISO 9000 standards; the second, with its epicenter in the United States, is national quality awards, largely with reference to total quality. The European emphasis on certification – and on mutual recognition of certification – clearly originated with the creation of the common market and the elimination of all barriers, including technical barriers, to the free circulation of goods. Attention to certification has since spread to countries with trading interests in Europe, most notably the United States. Although quality awards originated in Japan (back in 1951) and were subsequently introduced in Europe, their development has been most significant in America, where they are regarded as an effective tool for focusing the attention of the business community, the media, and government to the new global competitive strategy known as *total quality management* (TQM).

Although certification and awards have quite separate aims, they appear to share a common theme: assessment of company quality. But this apparent similarity needs to be carefully reexamined, since it is the source of distortions and unjustified extrapolations, such as the frequent extension of legitimate certification criteria to the total quality assessments typical of the main awards; or the use of examiners with the same professional background for both types of assessment.

These anomalies are hardly surprising. The western world's interest in quality has developed so swiftly, and at times so obsessively, that distortion and exploitation are inevitable. In our age of rapid change, the traditional organizational and managerial models no longer apply; the frantic search for new models and new competitive strategies has often focused attention on one or other of the more prominent elements of the new vision of total quality. The excessively high expectations this creates are soon disappointed, and the search begins for a new solution. This sort of situation is not uncommon: on one hand, managers are always anxious to find simple and easy-to-introduce formulas to solve the complex problems caused by an increasingly competitive marketplace and constantly changing points of reference; and on the other, countless ranks of "physicians" are always ready with a new medicine. Each new prescription becomes a buzzword and big business; but since by itself it cannot cure the disease, it is soon discarded and a new remedy is tried.

In the West, the huge potential of total quality risks being wasted, even before it has had an opportunity to demonstrate its worth. Signs of impatience and rejection are already emerging. Total quality – a global management approach – is blamed for failures whose real cause is faulty implementation, incomplete understanding, a short-term focus, and fragmented and inconsistent action.

To encourage a critical review of the tools used to promote total quality, and the way these tools are used, we need to take a fresh look at quality awards – another target of criticism in some countries – and at their central role in the total quality field today.

There is no doubt that national quality awards (and, today, international awards too) have played an important part in the development of a total quality culture. In particular, the MBNQA has been decisive in encouraging companies to adopt the practice of self-assessment in relation to a specific total quality model. This result is even more important than the Baldrige Award's primary goal, which is to arouse the interest of top management, foster a spirit of emulation among companies, and propose models of excellence.

The problems that arise with quality awards – problems that cannot be ignored if the awards are to maintain a positive role – are the reverse side of their merits: how to ensure their effectiveness in commending truly excellent companies on one hand and their value as self-assessment tools on the other. As far as the first problem is concerned, the objectives of the award must always be clearly stated: whether the focus is on results already achieved, or on improvement trends, or on the company quality system. The second problem – value as a self-assessment tool – is more critical. To a certain extent, the quality model and assessment criteria for an award are a matter of choice, as long as the organizing body clearly states the objectives of its award (then companies are free to accept these criteria and compete for the award, or not accept them and not compete). But the model and criteria for self-assessment should be objective, intended for general application, since improvement of the company is the sole aim. So, although differences between models are legitimate as far as awards are concerned, they are not legitimate at the level of self-assessment. Here, a certain consensus on a general TQM model and assessment criteria is desirable, even though each sector, and each company in each sector, will obviously have to personalize the model (and continue to adjust it to take account of changes in the business environment).

The principal aim of the IAQ national and international quality awards project is to provide those who have set up, are setting up, or plan to set up quality awards with a wide-ranging review of the current situation, together with historical information on the development of existing awards and, in the case of long-standing awards, a critical commentary on experience to date.

A second objective is to highlight the particular features of the TQM model on which each award is based, in order both to discover what the award aims to commend and to assess its specific features in the context of self-assessment. This final point will be of interest not only to those who are involved in quality awards, but also to quality managers and company managers in general. A comparative review of the various models and assessment approaches can help pinpoint the aspects that can be useful in personalizing a model to meet the needs of a particular company.

The papers of the academicians who have worked on this project follow this Introduction. They provide a highly detailed picture of the situation worldwide. The final paper, my own contribution, is a general analysis of quality awards and self-assessment, which takes company quality assessment as its starting point. The opinions it contains are of course subjective; they have already been expressed elsewhere, in particular during preparatory work on the European Quality Award, and hopefully will stimulate further debate.

My warmest thanks go to the following people for their contributions to this project: Y. Kondo, H. Kume, S. Shimizu, R. Wachniak, A. B. Godfrey, W. A. J. Golomski, K. E. Case, J. A. Goldsmith, R. Knowles, K. Jonson, A. Aune, A. Chauvel, J. A. Murphy, A. H. Zaludova, M. Bertin, and K. S. Stephens.

Tito Conti

Section I:

IAQ Project Papers

Chapter 1:

The Deming Prize

YOSHI KONDO

HITOSHI KUME

SHOICHI SCHIMIZU

History of the Deming Prize

It is well-known that the widespread adoption of companywide quality control (CWQC) or total quality control (TQC) based on statistical quality control (SQC) techniques in Japanese industry has resulted in major improvement in quality of the products and services proffered, together with much enhanced productivity and cost reduction.

For a number of years after World War II, the foremost task for Japan was to raise the standard of living through revitalization of the economy. For this, because of paucity of natural resources, there was no choice but to strive toward becoming a vigorous trading nation and at the same time to improve the image of pre-War, Japanese-made products. Nothing could have been more efficient and definite under the circumstances than the adoption and practice of SQC.

In one of the first steps of this approach, the Union of Japanese Scientists and Engineers (JUSE) invited W. Edwards Deming, American statistician and proponent of quality control (QC) techniques, in July 1950 to present a series of lectures at such seminars as the "8-Day Course on Quality Control" organized by JUSE. This provided vital stimulus for the early efforts in Japanese industry for the use of industrial QC.

The Deming Prize was instituted in 1951 by a formal resolution of the JUSE Board of Directors in grateful recognition of Deming's friendship and his achievements in the cause of industrial QC as proposed by the late Kenichi Koyanagi, a board member and one of the founders of JUSE.

Since that time, the adoption of QC and its techniques were seen in virtually every sector of Japanese industry, and there evolved in due time the concept of CWQC or TQC, which has come to attract much attention abroad.

It has become customary in Japan for corporations wishing to improve their performance in products or services to vie for the Deming Prize, not only for the prestige that goes with this honor but also for the benefit from the internal improvements that result from the complete implementation of CWQC or TQC that is needed in order to qualify at home and abroad.

Funding of the Deming Prize began with donation by Deming of the royalties received by him from the sale of the Japanese edition of his book *Some Theory of Sampling* and from his other works used or published in Japan, supplemented by donations from various other sources.[1] Today, the financing is undertaken by JUSE.

Until recently the Deming Prize was restricted to Japanese companies because its initial purpose was to encourage the development of QC in Japan. In recent years, however, strong interest in the Deming Prize has been shown by non-Japanese companies. The Deming Prize Committee has therefore revised the basic regulation, establishing in 1984 the "Regulations Regarding the Management of the Deming Prize" to allow the acceptance of overseas companies as candidates.

Categories

Under the Deming Prize Regulations, there are the Deming Prize for Individual Person, the Deming Application Prizes, and also the Quality Control Award for Factory adjudged by the Deming Prize Committee.

The Deming Prize for Individual Person

This prize is for a person who shows excellent achievement in the theory or application of SQC, or a person who makes an outstanding contribution to the dissemination of SQC.

The Deming Application Prizes

These prizes are for enterprises (including public institutions) or divisions that achieve the most distinctive improvement of performance through the implementation of CWQC based on SQC. Awarded especially to enterprises of medium or small size is the Deming Application Prize for Small Enterprise, while the prize awarded to corporate divisions is known as the Deming Application Prize for Division.

Organization of, and Examination and Selection by the Deming Prize Committee

For the examination, selection, and awarding of the Deming Prize, the Deming Prize regulations stipulate the establishment of the Deming Prize Committee, chaired by chairman of the Board of Directors of JUSE or a person recommended by the Board of Directors of JUSE.

The Deming Prize Committee consists of members commissioned by the chairman from men of learning and experience and officers of organizations related to QC.

Subcommittees are formed under the Deming Prize Committee, responsible for the selection of their respective Deming Prizes and other related matters. The Deming Prize Subcommittee consists of members drawn from university professors and QC experts in government and other nonprofit institutions. Personnel of profit-making business enterprises are excluded.

The administrative work of the Deming Prize Committee is carried out by JUSE in conformity to the Deming Prize regulations, with overall responsibility for the office work of the committee resting with the secretary general of JUSE.

The Deming Application Prizes are awarded each year by the Deming Prize Committee to the applicants adjudged meritorious as the result of strictly impartial examination and selection conducted by the committee and the subcommittee in charge.

Examination Processes

4.1 Examination Items

The manner in which such activities as investigation, research, development, design, purchase, production, inspection, sales, and so on which are essential for the proper control of product and service quality are conducted by each and every segment of the company is examined and judged. For example, each of the items listed here will be evaluated in regard to the method used to maintain the effective control over costs, profits, appointed dates of deliveries, safety, inventories, manufacturing processes, equipment maintenance, instrumentation, personnel and labor relations, education and training, new product development, research, relationship with subcontractors, associates, material suppliers and sales companies, handling of complaints, utilization of customers' opinions, quality assurance and after-sale services to customers, and relationship with companies to which products are delivered. The term *quality control* as used here denotes CWQC based on SQC techniques.

1. Company policy and planning
   How the policy for management, quality, and QC is determined and transmitted throughout all sectors of the company are examined together with the results being achieved. Whether the contents of the policy are appropriate and clearly presented are also examined.

2. Organization and its management
    Whether the scope of responsibility and authority is clearly defined, how cooperation is promoted among all departments, and how the organization is managed to carry out QC are examined.

3. QC education and dissemination
    How QC is taught and how employees are trained through training courses and routine work in the company concerned and in the related companies are examined. To what extent the concept of QC and statistical techniques are understood and utilized, and the activeness of QC circles are examined.

4. Collection, transmission, and utilization of information on quality
    How the collection and dissemination of information on quality within and outside the company are conducted by and among the head office, factories, branches, sales offices, and the organizational units are examined, together with the evaluation of the organization and the systems used. How fast information is transmitted, sorted, analyzed, and utilized are also examined.

5. Analysis
    Whether the critical problems regarding quality are properly grasped and analyzed with respect to overall quality and the existing processes, and whether the results are being interpreted in the frame of the available technology are subject to scrutiny, while checks are conducted on whether proper statistical methods are being used.

6. Standardization
    The establishment, revision, and rescission of standards and the manner of their control and systematization are examined, together with the use of standards for the enhancement of company technology.

7. Control ("Kanri")
    How the procedures used for the maintenance and improvements of quality are reviewed when necessary are examined. Also scrutinized are how the responsibility for and the authorities over these matters are defined, while checks are conducted on the use of control charts and other related statistical techniques.

8. Quality assurance

    New product development, quality analysis, design, production, inspection, equipment maintenance, purchasing, sales, services, and other activities at each stage of the operations, which are essential for quality assurance, including reliability, are closely examined, together with the overall quality assurance management system.

9. Effects

    What effects were produced or are being produced on the quality of products and services through the implementation of QC are examined. Whether products of sufficiently good quality are being manufactured and sold are examined. Whether products have been improved from the viewpoint of quality, quantity, and cost, and whether the whole company has been improved not only in the numerical effect of quality and profit, but also in the scientific way of thinking of employers and employees and their heightened will to work are examined.

10. Future plans

    Whether the strong and weak points in the present situation are properly recognized and whether the promotion of QC is planned in the future and is likely to continue are examined.

4.2 The Description of QC Practices

After submitting the application form and receiving the acceptance notification from the Deming Prize Committee, the applicant is required to submit a description of QC practices together with a company business prospectus. The description of QC practices should present in concise and concrete terms the actual state of the QC currently practiced.

4.2.1 Dividing of the Description of QC Practices

It is suggested that the description is subdivided, as a rule, into the following descriptive divisions, each under separate cover: the company as a whole, the head office, the divisions, the branches, the factories, the research facilities, and the sales offices.

No description needs to be submitted for divisions employing less than 10 employees. On-site examination does not necessarily follow this dividing of the description of QC practices.

4.2.2 Contents

In accordance with the following instructions, the actual conditions under which QC is being implemented should be concisely described in line with items to be examined. In particular, the relationship between cross-functions and the departments should be clarified with the explanation as to how responsibilities are organizationally allocated and collaborative coordination is maintained for such activities as new product development, quality assurance, volumetric control, and control of costs.

In the case of entries for the Deming Application Prize for Divisions, the relationship between the division concerned and the head office in regard to such functions as quality planning and quality assurance should be clearly explained. In this case the description of the division concerned should be submitted instead of that of the company as a whole and its head office.

The description of QC practices should not be too formal when describing about the examination items, but should give a concise explanation of the company's priorities, features, weak points, and future plans with regard to QC so that it may be useful for the on-site inspection.

Instructions

The style of writing and the arrangement of the examination items may be at the applicant's discretion, but it is necessary that the following points are included.

1. The features and general outline of the descriptive divisions, and a summary of products

2. The relationship between the policies of the company and the head of the divisions, and the policies for QC (short- and long-term)

3. The relationship between the organization of the company and the divisions, and the organization for QC

4. The reasons for the introduction of QC

5. The progress made in QC activities (the changes undergone in the way QC has been managed), the changes made in the policies for QC and the current QC policies

6. The features of the QC activities (activities to which priorities are given)

7. The quality level of main products (compared with that of competitors'), self-imposed evaluation of the products and the reactions to the results

8. Tangible and intangible results of the implementation of QC

9. Those aspects of QC which have not been fully achieved, and plans for improving them

10. Other matters which may be helpful to on-site inspection (points of special importance, examples of the implementation of QC and so on)

It is necessary that the description of QC practices of the head office includes the organization of the head office and the activities of each division of the head office.

It is necessary that the description of QC practices of the whole company includes the following points.

1. The summarization of activities within the whole company according to functions, such as quality assurance, the control of the date of delivery, and so on

2. A list of directors, together with an account of the scope of their responsibilities, their opinions of QC and participation in QC activities, described and prepared by themselves, which may be helpful for the interview with executive officers of the company

4.2.3 Number of Copies

Thirty copies should be submitted for each descriptive division, together with the same number of copies of a list of descriptive divisions. If the number of descriptive divisions is so large, the secretariat of the Deming Prize Committee should be consulted.

4.2.4 Format of the Description of QC Practices

– *Page size:* A4 (210 x 297 mm) or similar size

– *Prescribed form:* To be written in Japanese from left to right, and bound at the left end. If necessary, an English version of the description of QC practices may be attached, in single-spaced typewritten form.

– *Number of pages:* The standard number of pages including the charts is given in Table 1.1. Each page shall not contain more than 1000 characters, and a double-sized page is considered as two pages.

Table 1.1. The number of pages allowed for the description of QC practices.

Number of employees in the division described	Maximum number of pages
100 or less	50 pages
1000 or less	60 pages
2000 or less	75 pages
for each additional 500 employees	5 pages added

If the head office and factories must be combined in the description, another 50 pages may be added. For the description of the company as a whole, 50 pages are allowed.

Instructions

A slight number of pages may be added to the standard maximum as pre-scribed in Table 1.1, if necessary. Charts including small characters and the use of larger folded pages should be avoided.

4.2.5 Submission of the Description of QC Practices

The applicant is requested to submit 20 copies of the description of QC practices, the company prospectus, and the documents mentioned here, not later than March 31.

The company prospectus should contain the history of the company, particulars of the products manufactured and sold, descriptions covering the size of each division, documents showing the management (for example, financial statements), and other informative documents (brochures, catalogs, technical term glossaries, and so on).

Also required are: a list of holidays and dates unsuitable for on-site inspection; rosters of personnel employed at each of the divisions to be inspected (see 4.3.2.2); names, office and home addresses, telephone numbers, facsimile and telex numbers, cable addresses, and so on of the contact personnel for the company and for each of the divisions to be visited; and information on the nearest airport, and the travel time required between airport and hotel, hotel and inspection site, and so on).

The documentation papers for each division should be assigned a serial number and be contained as a set in a large envelope bearing the same number.

Notify the secretariat in advance of the planned date of the submission of the description.

4.2.6   Handling the Description of QC Practices

Confidential matters in the description to be submitted for examination should be clearly indicated. The applicant, because of confidentiality requirements, may refuse the explanation, furnishing of materials, or access to work site. When such refusal is of such extent that the examination of the applicant's QC performance is seriously impeded, the committee may decide to terminate the examination. In regard to information indicated as confidential, every possible precaution is taken to ensure that there is no disclosure to third parties.

4.3   Examination Objective and Procedure

4.3.1   Examination Objectives

The main objective of the examination is to ascertain that good results are being comprehensively achieved through the implementation of CWQC, particularly in regard to the potential for the future advancement of the company's CWQC. Especially, emphasis is laid on whether the QC being practiced is of the type most suitable for the business, its size, and other conditions of the company.

The practice of CWQC in this case is defined as the designing, production, and supply of products or services of a quality level demanded by the customer at an economically acceptable cost, the basic approach being of customer satisfaction together with a wide attention to public welfare. Also implied are better understanding and their application of statistical concepts and methods by all members of the company in the train of activities involving investigations, research, development, designing, purchasing, manufacturing, inspection, and sales as well as other related activities inside and outside the company, together with the rational reiteration of planning, implementation, evaluation, and improvement, all for the attainment of beneficial business goals.

New product development, administration of research and development (R&D), control of materials and supplies, management of physical facilities, instrumentation control, management of subcontracted work, personnel education and training, and various other activities will also be subject to examination from the viewpoint of quality assurance.

4.3.2 Examination Procedure

Investigation work relating to examination is undertaken by the Deming Application Prize Subcommittee, and the result is reported to the chairman of the Deming Prize Committee. The investigation work consists of the examination of documents and the on-site inspection.

4.3.2.1 Examination of Documents

The applicant's document is examined and judged to ascertain whether QC is being practiced systematically and effectively throughout the company. Upon passing this examination, the applicant is duly notified, and on-site inspection is performed in the manner set forth here, on the basis of the information contained in the description of QC practices. If the applicant is disqualified after the examination of documents, a written opinion giving the reasons for nonacceptance will be provided.

4.3.2.2 Inspection Units

Inspection units are the chief executive officers (CEOs), the head and regional offices, corporate divisions, branches, works and factories, laboratories, and sales and local offices. (Workplaces with less than 10 employees are not counted as units.)

Not all the units cited by the applicant are inspected. Visited are only the units selected by the committee, in some cases a single location being designated depending on the actual conditions.

These matters are subject to consultations between the applicant and the committee.

4.3.2.2 On-Site Inspection

On-site inspection is performed at the units selected by the committee. However, as mentioned in 4.3.2.2, inspection by sampling may be conducted if deemed necessary. Also, the inspection units need not necessarily be the same as the division set forth in the description of QC practices.

The examiners, as a rule, consist of more than two persons per inspection unit (normally, two to six persons), one of whom is designated as the team leader, in charge of liaison and consultation with the applicant.

The schedule of the on-site inspection is notified to the applicant after the entire schedule is finalized.

The names of the examiners assigned to each inspection unit are reported to the applicant at the earliest possible time, at least two months ahead of the scheduled date. However, the number of examiners may be changed.

On-site inspection is conducted with the least possible reliance on explanations by the applicant, stress being laid on questioning and investigation by the examiners. The days to be spent at the inspection units are generally as listed in Table 1.2.

Table 1.2. Typical days spent at inspection units.

Inspection unit	Days required
Head office, regional offices	2 or 3 days
Factories	2 or 3 days
Branches, sales offices	1 or 2 days
Laboratories	1 or 2 days
When the head office and a factory are integrated	2 or 3 days

Unusual circumstances reported by the applicant will be given consideration, and the committee will decide the suitable duration of visits.

On-site inspection consists of three parts: Schedule A, Schedule B, and interviews with the CEO.

1. Schedule A is to be prepared by the applicant and decided after consultations with the inspection team leader concerned. Schedule A is made up of two parts: the presentation of important points and the presentation of operations sites.

   In the case of places of business, the time allocation to these two presentations should be roughly 2:1.

   Presentation of important points

   a. Explanation of the points considered to be particularly important in the description of QC practices

   b. Explanation of the situation after the submission of the description of QC practices

   c. Discussion of the contents of these reports (as a rule, 15 minutes given out of each hour)

   Materials in support of the presentation of important points may be submitted at the time of presentation or distributed to the examiners, who are under no obligation to read through all such materials.

   Presentation of operations sites

   In the case of a production factory, a general explanation of the processes and the methods of management of manufacturing, inspection, testing, packaging, storage, and so on should be given. The presentation of other sites should be structured generally in the same way.

2. Schedule B consists of on-site inspection and general interrogation. After the examination of Schedule A, the examiners with due respect for the applicant's opinions prepare a draft for the implementation of Schedule B, which becomes final upon approval by the applicant.

   The on-site inspection as prescribed by Schedule B is performed as directed by the examiners: the methods and procedures all being as determined by the examiners. For instance, it may be that on-site inspection will entail only simple questioning. If, at the time of on-site inspection, confusion or embarrassment is caused on the part of the applicant, the inspection team leader should be promptly informed and consulted.

3. For the interview with the CEO, it may be that the inspection team leader concerned will consult with the applicant beforehand as to who will participate on the applicant's side.

4. On-site inspection is conducted, as a general rule, from 9:00 A.M. to 5:00 P.M. with an hour's break for lunch. Further breaks of 15 minutes each will be taken during the morning and afternoon sessions. All these intermissions will be counted as time of inspection work.

5. Examples of on-site inspection hours are shown in Table 1.3. In the case of the head office, two additional hours are spent for the interview with the CEO.

Table 1.3. Examples of on-site inspection hours.

Site	Days needed	Schedule A	Schedule B
Work site	1	3 hours, A.M.	4 hours, P.M.
Work site	2	6 hours, A.M.	8 hours, P.M.
Work site	3	9 hours, A.M.	12 hours, P.M.
Head office	1	3 hours, A.M.	3 hours, P.M.
Head office	2 (1st day)	5 hours, A.M.	2 hours, P.M.
	(2nd day)		5 hours, P.M.
Head office	3	9 hours, A.M.	10 hours, P.M.

6. At the inspection site, the explanatory materials showing the operations at each stage of the production flow should be available, together with the data used for routine control of quality and production. There is no need to prepare special wall charts and other materials for the investigation unless requested.

7. As need arises, the on-site inspection is conducted by teams consisting of two or more members, these teams being on separate assignments. Also, each member of a single team may be undertaking dissimilar work.

8. When on inspection, the team members may wish to conduct private talks among themselves. The applicant is requested to provide facilities for such purposes.

9. Responses to the inspection team interrogations, unless otherwise indicated by the examiners, are restricted to the personnel employed at the unit under inspection. If somebody outside the unit is required to answer a question, the inspection team leader should be consulted.

10. While on-site inspections are in progress, no advice or guidance will be provided as a rule. Neither will there be any comment on the results after the inspection.

11. The on-site inspection is conducted in Japanese. If necessary, the applicant may furnish two or more Japanese language interpreters.

12. Companies related to the quality assurance achievement of the applicant (particularly those in the same business group, subcontractors, vendors, consignees, distributors, and so on) may be investigated, not for the evaluation of their respective companies but for data providing a basis for reference in assessing the applicant. If deemed necessary by the applicant, such investigations may be included in Schedule A, while when the examiners so decide, inspection may be included in Schedule B, subject to the consent of the company involved. The investigations of reference purposes are not always taken in each inspection limit.

Evaluation of Inspection Results

5.1 Evaluation

5.1.1 Scoring Method

The examination committee evaluates the results reported by the on-site inspection teams comprehensively. Each team member's score is treated equally, on the basis of 100 points.

The passing points are

1. The CEO . . . . 70 points or more

2. Whole company average, excluding the CEO . . . . . 70 points or more

3. Minimum for any inspected unit . . . . 50 points or more

5.1.2 Totalization

1. The score for each inspected unit is the median value of the points awarded by each of the inspection team members.

2. For the head office as an inspection unit, the scoring for the CEO and other organizational departments is done separately.

3. The score for the whole company is in terms of the weighted average of the scores won by each inspected unit, apart from the CEO.

4. There is no disclosure of these scores.

5.2 Judgment

The judgment of the Deming Prize Committee is based on the report to it from the Deming Application Prize Subcommittee.

1. The Deming Application Prize is awarded by the Deming Prize Committee to the applicants considered to have qualified on the basis of the aforementioned report.

2. In the event the passing point score has not been attained by the applicant, judgment is reserved, and unless withdrawal is expressed by the applicant, the examination process is not terminated. Extension is limited to two more times and to a period of three years. In the case of extended

examinations, emphasis is laid on the controversial points raised at the time of the preceding examination. The applicant is selected as an awardee when it is considered to attain the passing standard after the points of issue are improved.

Note

1. W. Edwards Deming, *Some Theory of Sampling.*

Chapter 2:

American Quality Awards: Profiles In Excellence

RAYMOND WACHNIAK

*"Quality is not an art,
it is a habit"*

Aristotle

Despite some criticism, American quality awards are positioned exactly where they should be – as an agent for transforming U.S. business.

Before the awards came into being there was the need to address quality at a national level. In the United States a renewed dedication to the proposition that America could continue to benefit by cultivating the habit of never-ending quality improvement got a kickstart by the congress and President Ronald Reagan in the early 1980s. It was in these early years that the seeds of national quality awards were planted.

The purpose of this chapter is to guide you through several American quality awards. They are the Malcolm Baldrige National Quality Award (MBNQA), which is for private sector business and industry; the (U.S.) Presidential Award for Quality and the Quality Improvement Prototype (QIP) Award, which are for federal government agencies and organizations; the U.S. National Aeronautics and Space Administration (NASA) Excellence Award for Quality and Productivity, which is given to its contractors, subcontractors, and suppliers; and the Institute of Industrial Engineers (IIE) Award for Excellence in Productivity Improvement for companies and organizations.

Malcolm Baldrige National Quality Award

In 1987, for the first time, the United States claimed a quality prize of its own. The Baldrige Award – a shortened version of the official designation – is now recognized by quality experts the world over. The Baldrige Award has reshaped managers' thinking and behavior. It codifies the principles of quality management in clear and accessible language. It provides a fresh perspective on such acknowledged goals of customer satisfaction and increased employee involvement.

"Just as important," said former President Reagan, in 1988, "it offers a vehicle for companies, large and small, in manufacturing and in services, to examine their own approaches to quality. It offers companies a standard with which to compare their own progress to that of the country's very best."

The Baldrige Award is presented annually to companies that exhibit excellence in all aspects of quality. A maximum of two awards may be presented in each of three categories: large manufacturing, service, and small business. The award questions span seven categories: leadership, information and analysis, planning, human resources, product assurance, quality results, and customer satisfaction. Chapter 3 provides a more comprehensive report on the Baldrige Award.

Presidential Award for Quality and the Quality Improvement Prototype Award

Since 1988, federal government agencies and organizations have had an opportunity to be recognized and honored for their improvement efforts. Administered by the Federal Quality Institute, the Presidential Award for Quality and the QIP Award are helping to build and sustain the growing momentum evidenced at the federal level for excellence in product and service quality.

There are similarities between the two awards, but qualification for each differs. Additionally, the Presidential Award recipients must show an improvement trend in a three- to five-year range, while those of the QIP Award reflect a one- to two-year horizon. A comparison of their application/evaluation procedure and criteria are also detailed in chapter 4.

The George M. Low Trophy – NASA Excellence Award for Quality and Productivity

This award was named for George M. Low, a former NASA deputy administrator whose contributions to the U.S. space program exemplified a quality philosophy that was far ahead of its time.

The NASA Excellence Award acknowledges the pivotal role current NASA contractors, subcontractors, and suppliers play in meeting the demands of the nation's space program. Awardees must demonstrate sustained excellence and outstanding achievements in quality and productivity for three or more years.

The objectives of the award are to

- Increase public awareness of the importance of quality and productivity to the nation's aerospace program and industry in general

- Encourage domestic business to continue efforts to enhance quality, increase productivity, and thereby strengthen competitiveness

- Provide the means for sharing the successful methods and techniques used by the applicants with other American enterprises

A more detailed description of this award can be found in chapter 5.

IIE Award for Excellence in Productivity Improvement

IIE is an example of a professional society's contribution to quality revolutions taking place in the United States. The IIE Award for Excellence in Productivity Improvement recognizes companies and organizations which, through diligent and innovative means, have accomplished significant, measurable, and observable achievements which increased productivity, eliminated human drudgery, and improved quality.

Participating companies benefit from the self-assessment required during the application process. Finalists receive a detailed feedback report prepared by the awards committee. This annual award is open to all companies and organizations, including subsidiaries or divisions within a company or organization, throughout the world. Learn more about this award by reading the detailed description in chapter 6.

Winners Shaping and Sharpening American Management Practices

Built into these award processes is an understanding of the requirements for quality excellence and a sharing of information on successful quality strategies. From previous winners we have learned of six-sigma quality, total cycle time reduction, pulse points, benchmarking, LUTI, and many others.

From Motorola we learned of six sigma quality, that is, designing products that will accept reasonable variation in component parts, and developing manufacturing processes that will produce a minimum variation in the final output product. This translates to a defect level of 3.4 parts per million (ppm) across a company, services as well as products. A second practice is the reduction in total cycle time, defined as the elapsed time from the moment a customer places an order to the time it is delivered. In the case of a new product it is the time from conception to shipping.

From Westinghouse Electric Corporation, Commercial Nuclear Fuel Division (CNFD), the practice of pulse points came in use. CNFD's fulfillment of its yearly documented quality plan requires accurate, ongoing measurement of performance. These objectives are called *pulse points*. They are represented as areas on a symbolic human figure. Using pulse points underscores the fact that it is imperative to perform control checks and progress against objectives, much like a human patient is monitored for vital signs. Pulse points are converted to individual plans that cascade from managing director to every employee. It is the golden thread that ties the organization together.

From IBM Rochester we learned the art of benchmarking. The company's initial strategy in 1979 was cost benchmarking. Today benchmarking is defined as a continuous process of measuring products, services, and practices against the toughest competitors or organizations recognized as leaders. This 10-step process enables organizations "to move from an inward, arrogant focus to an outward, critical, self-assessing style." Little wonder that benchmarking organizations and committees are forming throughout the country.

From Xerox Corporation Business Products and Systems we learned of many enhancement practices. One such practice is "LUTI," the need to ensure that all levels of management learn, understand, and apply total management concepts. All managers are required to *Learn* about TQM by attending (how to) training. *Understanding* is assured by a thorough assessment process. Managers are then required to *Teach* their subordinates, and assure students' understanding by *Inspecting* their work. Hence, the acronym LUTI.

Literally hundreds of good practices have come to light as a result of these national awards. Only a few have been reported here.

Results of a GAO Study

In May 1992, at the request of 29 members of the U.S. Congress, the U.S. General Accounting Office (GAO) studied and reported on U.S. companies' improved performance through quality efforts. Here are some of the GAO's results.

GAO's review of 20 companies that were among the highest scoring applicants in 1988 and 1989 for the MBNQA indicated the following.

- Companies that adopted quality management practices experienced overall improvement in corporate performance. In nearly all cases, companies that used TQM practices achieved better employee relations, higher productivity, greater customer satisfaction, increased market share, and improved profitability.

- Each of the companies studied developed its practices in a unique environment with its own opportunities and problems. However, there were common features in their quality management systems that were major contributing factors to improved performance.

- Many different kinds of companies benefited from putting specific management practices in place. However, none of these companies reaped those benefits immediately. Allowing sufficient time for results to be achieved was as important as initiating a quality management program.

Chapter 3:

The Malcolm Baldrige National Quality Award: Seven Years of Progress, 7000 Lessons Learned

A. BLANTON GODFREY

The Origin of the MBNQA

During the 1980s there was a growing interest in the United States in promoting what is now called total quality management (TQM). Many leaders in the United States felt that a national quality award, similar to the JUSE Deming Prize, would help stimulate the quality efforts of U.S. companies.

A number of individuals and organizations proposed such an award, leading to a series of hearings before the U.S. House of Representatives Subcommittee on Science, Research, and Technology. Finally on January 6, 1987, the Malcolm Baldrige National Quality Improvement Act of 1987 was passed. The act was signed by President Ronald Reagan on August 20, 1987 and became Public Law 100-107. This act provided for the establishment of the MBNQA program. The purpose of this award program was to help improve quality and productivity by

1. Helping stimulate American companies to improve quality and productivity for the pride of recognition while obtaining a competitive edge through increased profits

2. Recognizing the achievements of those companies which improve the quality of their goods and services and providing an example to others

3. Establishing guidelines and criteria that can be used by business, industrial, governmental, and other organizations in evaluating their own quality improvement efforts

4. Providing specific guidance for other American organizations that wish to learn how to manage for high quality by making available detailed information on how winning organizations were able to change their culture and achieve eminence

The act provided that up to two awards could be presented to companies in each of three categories: small businesses, companies or their subsidiaries, and companies which primarily provide services. The act also stated that companies must apply for the award by submitting an application, in writing, for the award. And the company must permit a rigorous evaluation of the way in which the business and other operations have contributed to improvements in the quality of goods and services.

The act also called on the director of the National Bureau of Standards [now the National Institute of Standards and Technology (NIST)] to rely upon an intensive evaluation by a competent Board of Examiners which shall review the evidence submitted by the organization and, through a site visit, verify the accuracy of the quality improvements claimed. The examination should encompass all aspects of the organization's current quality management in its future goals. The award shall be given to organizations which have made outstanding improvements in the quality of their goods or services (or both) and which demonstrate effective quality management through the training and involvement of all levels of personnel in quality improvement.

In addition to the establishment of the Board of Examiners, the act also called for the establishment of a Board of Overseers consisting of at least five individuals who have demonstrated preeminence in the field of quality management.

Description of the MBNQA

In creating the MBNQA, the first step was to develop the criteria which would be used to evaluate the organizations applying. The director of the National Bureau of Standards selected Curt W. Reimann as director of the MBNQA. Reimann immediately began calling on individuals and organizations throughout the United States and world for their suggestions and contributions to creating the criteria and the process by which these criteria would be evaluated. Reimann and his staff collected much information on other awards, such as the JUSE Deming Prize and the NASA Excellence Award, as background information.

They then selected a small team of volunteers to help create the first draft of the criteria. These draft criteria were then reviewed in intensive focus group sessions by selected experts from organizations throughout the United States. One of the most important actions taken by the director, his team, and the volunteers at this stage was to create a clear design strategy for the award program. The elements of the strategy were

- To create a national value system for quality
- To provide a basis for diagnosis and information transfer
- To create a vehicle for cooperation across organizations
- To provide for a dynamic award system which would evolve through consensus and to be continuously improved

The design strategy has been followed carefully. The award criteria have been changed and improved each year. For more information about current award criteria and an application form, contact NIST.[1]

The MBNQA criteria are the basis for making awards and giving feedback to the applicants. The criteria also have three other important purposes.

- They help raise quality performance standards and expectations.

- They facilitate communication and sharing among and within organizations of all types based on a common understanding of key quality and operational performance requirements.

- They serve as a working tool for planning, training, assessment, and other uses.

Core Values and Concepts

There are 10 core values and concepts embodied in an award criteria. These core values and concepts are

1. **Customer-Driven Quality.** Emphasis here is placed on product and service attributes that contribute value to the customer and lead to customer satisfaction and preference. The concept goes beyond just meeting basic customer requirements and also includes those that enhance the product and service attributes and differentiate them from competing offerings. Customer-driven quality is thus described as a strategic concept directed toward customer retention and market share gain.

   This focus on the customer was emphasized by George Bush in 1992 when he said that, in business, there is only one definition of quality – the customer's definition. With the fierce competition of the international market, quality means survival.[2]

2. **Leadership.** A key part of the MBNQA focus is on senior executive leadership. The leaders must create a customer orientation, clear and visible quality values, and high expectations. This concept stresses the personal involvement required of leaders. This involvement extends to areas of public responsibility and corporate citizenship as well as to areas of development of the entire workforce. This concept also emphasizes such activities as planning, communications, review of company quality performance, recognition, and serving as a role model.

3. **Continuous Improvement.** This concept includes both incremental and breakthrough improvement activities in every operation, function, and work process in the company. It stresses that improvements may be made through enhancing value to customers; reducing errors, defects, and waste; improving responsiveness and cycle time performance; improving productivity and effectiveness in the use of all resources; and improving the company's performance and leadership position in fulfilling its public responsibilities and corporate citizenship.

4. **Employee Participation and Development.** This concept stresses the close relationship between employee satisfaction and customer satisfaction. It explains the value of employee satisfaction measurement and how this is an important indicator of the organization's overall performance. There is an increasing awareness in the United States that overall organization performance depends increasingly on workforce quality and involvement. Factors that bear upon the safety, health, well-being, and morale of employees need to be part of the company's continuous improvement objectives.

5. **Fast Response.** The value of shortening time cycles is also emphasized. Faster and more flexible response to customers is becoming each year a more critical requirement of business management. Improvements in these areas often require redesigning work processes, eliminating unnecessary work steps, and making better use of technology. Measures of time performance should be among the quality indicators used by leading organizations.

6. **Design Quality and Prevention.** Throughout the criteria the importance of prevention-based quality systems is highlighted. Design quality is a primary driver of downstream quality. This concept includes fault-tolerant (robust) products and processes. It also includes concept-to-customer times, the entire time for the design, development, production, and delivery to customer of new goods and services.

   The concept of continuous improvement and corrective action involving upstream interventions is also covered here. This concept stresses that changes should be made as far upstream as possible for the greatest savings.

7. **Long-Range Outlook.** This concept stresses the need to take a long-range view of the organization's future and consider all stakeholders: customers, employees, stockholders, and the community. Planning must take into

account new technologies, the changing needs of customers and the changing customer mix, new regulatory requirements, community/societal expectations, and competitors' strategies. Emphasis is also placed on long-term development of employees and suppliers and on fulfilling public responsibilities and serving as a corporate citizenship role model.

8. **Management by Fact.** This concept stresses the need to make decisions based on reliable data, information, and analyses. These data need to accurately reflect the needs, wants, expectations, and perceptions of the customers; to give accurate descriptions of the performance of goods and services sold; to reflect clearly the market situation; to portray accurately the offerings, performance levels, and satisfaction levels of competitors' goods and services; to provide clear findings of employee-rated issues; and to accurately portray the cost and financial matters. The role of analysis is stressed. Here, also, emphasis is placed on the role of benchmarking in comparing organizational quality performance with competitors' or best-in-class organizations' performance.

   The need for organizationwide performance indicators is also stressed. These indicators are measurable characteristics of goods, services, processes, and company operations. They are used to evaluate, track, and improve performance. They should be clearly linked to show the relationships between strategic goals and all activities of the company.

9. **Partnership Development.** The need to develop both internal and external partnerships to accomplish overall goals is also emphasized. These partnerships may include labor-management relationships; relationships with key suppliers; working agreements with technical colleges, community colleges, and universities; and strategic alliances with other organizations.

10. **Corporate Responsibility and Citizenship.** The core values and concepts also emphasize that the organization's quality system should address corporate responsibility and citizenship. This includes business ethics, protection of public health, public safety, and the environment. The company's day-to-day operations and the entire life cycle of the products sold should be considered as they impact health, safety, and environment. Quality planning should anticipate any adverse impacts from facilities management, production, distribution, transportation, use, and disposal of products.

Corporate responsibility also refers to leadership and support of such areas as education, resource conservation, community services, improving industry and business practices, and sharing of nonproprietary, quality-related information, tools, and concepts.

The Criteria

The core values and concepts described earlier are embodied in seven categories.

1. Leadership

2. Information and Analysis

3. Strategic Quality Planning

4. Human Resource Development and Management

5. Management of Process Quality

6. Quality and Operational Results

7. Customer Focus and Satisfaction

The dynamic relationships among these seven categories are best described in Figure 3.1, which illustrates four basic elements: the driver, the system, the measures of progress, and the goals. The seven categories are further subdivided into 28 examination items and 92 areas to address. The seven categories, the 28 examination items, and the points for each category and examination item are shown in Figure 3.2.

The areas to address give specific instructions as to what information should be contained in the application form. Notes supporting each section give further explanation and clarification. The notes also help the applicant understand where certain data should be reported when there are several seeming possibilities.

An example of item 2.2 and its four areas to address is provided in Figure 3.3.

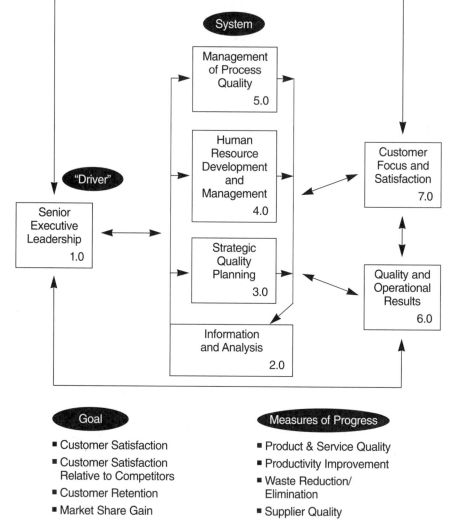

Source: National Institute of Standards and Technology, 1993 Awards Criteria,
Malcolm Baldrige National Quality Award, Gaithersburg, Md., 1992.

Figure 3.1.   Baldrige Award criteria framework: dynamic relationships.

1.0 Leadership (95 points)
1.1 Senior Executive Leadership (45)
1.2 Management for Quality (25)
1.3 Public Responsibility and Corporate Citizenship (25)

2.0 Information and Analysis (75 points)
2.1 Scope and Management of Quality and Performance Data and Information (15)
2.2 Competitive Comparisons and Benchmarking (20)
2.3 Analysis and Uses of Company-Level Data (40)

3.0 Strategic Quality and Company Performance (60 points)
3.1 Strategic Quality and Company Performance Planning Process (35)
3.2 Quality and Performance (25)

4.0 Human Resource Development and Management (150 points)
4.1 Human Resource Planning and Management (20)
4.2 Employee Involvement (40)
4.3 Employee Education and Training (40)
4.4 Employee Performance and Recognition (25)
4.5 Employee Well-Being and Satisfaction (25)

5.0 Management of Process Quality (140 points)
5.1 Design and Introduction of Quality Products and Services (40)
5.2 Process Management: Product and Service Production Delivery Processes (35)
5.3 Process Management: Business Process and Support Services (30)
5.4 Supplier Quality (20)
5.5 Quality Assessment (15)

6.0 Quality and Operational Results (180 points)
6.1 Product and Service Quality Results (70)
6.2 Company Operational Results (50)
6.3 Business Process and Support Service Results (25)
6.4 Supplier Quality Results (35)

7.0 Customer Focus and Satisfaction (300 points)
7.1 Customer Expectations: Current and Future (35)
7.2 Customer Relationship Management (65)
7.3 Commitment to Customers (15)
7.4 Customer Satisfaction Determination (30)
7.5 Customer Satisfaction Results (85)
7.6 Customer Satisfaction Comparison (70)

Total Points: 1000

Source: National Institute of Standards and Technology, 1993 Awards Criteria, *Malcolm Baldrige National Quality Award, Gaithersburg, Md., 1992.*

Figure 3.2. MBNQA 1993 examination items and point values.

Results

One of the frequent criticisms of the MBNQA is that too much emphasis is placed on quality systems and too little is placed on quality results. Tito Conti describes the strengths and weaknesses of the Deming Prize, the European Quality Award, and the MBNQA in what J. M. Juran calls "a most incisive paper."[3] Conti's criticisms of systems-based assessments are right on the mark. The proof of the effectiveness of any quality system *must* be in the results produced by the system.

2.2. Competitive Comparisons and Benchmarking *(20 pts)* Describe the company's processes, current sources and scope, and uses of competitive comparisons and benchmarking information and data to support improvements of quality and overall company operational performance. ☑ Approach ☑ Deployment ☐ Results	**AREAS TO ADDRESS** **1.** How the company uses competitive comparisons and benchmarking information and data to help drive improvement of quality and company operational performance. Describe: (1) how needs are determined; and (2) criteria for seeking appropriate comparison and benchmakring information – from within and outside the company's industry. **2.** Brief summary of current scope, sources, and principal uses of each type of competitive and benchmark information and data. Include: (1) customer-related; (2) product and service quality; (3) internal operations and performance, including business processes, support services, and employee-related; and (4) supplier performance. **3.** How competitive and benchmarking information and data are used to improve understanding of processes, to encourage breakthrough approaches, and to set "stretch" objectives. **4.** How the company evaluates and improves its overall processes for selecting and using competitive comparisons and benchmarking information and data to improve planning and company operations.

Source: National Institute of Standards and Technology, 1993 Awards Criteria, *Malcolm Baldrige National Quality Award, Gaithersburg, Md., 1992.*

Figure 3.3. An example: item 2.2 and its four areas to address.

Conti points out the MBNQA's apparent overfocus on systems and underfocus on actual results. In actual fact, the applicants for the MBNQA have always emphasized results, sometimes even entering charts and data in inappropriate places in the application form. The examiners also looked for results in almost every area addressed.

However, the language in the application form was not clear in the early years, and it was possible to interpret the application process as only giving 10 percent weight to customer satisfaction results and 10 percent weight to internal results. Some companies, and many reviewers, read the guidelines this way.

The 1993 application guidelines have attempted to correct some of this misunderstanding. Beside each area to address is a small chart indicating whether approach, deployment, or results are to be described. It is now clear that 335 points out of 1000 are only possible through results; none of these points are given for approach or deployment. Other areas to be addressed, totaling 130 points, also include results as part of the assessment. In fact, other areas also imply results for many applicants.

For example, examination item 2.3, "Analysis and Uses of Company-Level Data," does not specifically ask for or score for results. Yet, in the explanatory note for area 2.3a, it states

> Analysis appropriate for inclusion in 2.3a could include relationships between and among the following: the company's product and service quality improvement and key customer indicators such as customer satisfaction, customer retention, and market share; the relationship between customer relationship management strategies and changes in customer satisfaction, customer retention, and market share; cross comparisons of data from complaints, post-transaction follow-up, and won/lost analyses to identify improvement priorities; the relationship between employee satisfaction and customer satisfaction; cost/revenue implications or customer-related problems; and, rates of improvement in customer indicators.[4]

Not surprisingly, many companies find the easiest way to address this area is to give specific examples of what data they collect and what analyses they do and to show the changes in the results over time.

Other examples abound in the application guidelines. In examination item 5.5, "Quality Assessment," area to address 5.5b states clearly:

> how assessment findings are used to improve: products and services; processes; practices; and supplier requirements. Describe how the company verifies that assessment findings lead to action and that the actions are effective.[5]

Again, applicants often use examples to describe how assessment findings were used and what change occurred. These examples describe results of the quality assessment process. This is how the company demonstrates that the actions are effective.

The actual applications are full of charts, graphs, tables, and other forms of results. The winning companies are well on the way to management by fact, and it is not surprising that they report their activities in fact-rich documents. The examiners expect this, and often refuse to score any examination item highly that doesn't have convincing data to support a statement. One of the most common statements on a scored application is "Lack of evidence to support claim of . . . ."

Another misconception about the scoring is a belief that the examiners and judges rely wholly on a total score in making their final decisions on applicants. This is not at all the case. The seven category scores are always highly visible to all examiners and judges, and individual category scores are discussed at length. It is highly unlikely that a company scoring poorly in any single category would ever be selected for an award. In the almost-impossible-case-to-imagine of a company having perfect scores in all categories except those solely judged on results (a possible score of 665), the company would probably still not even be selected for a site visit much less an award. Even with 50 percent scores in the results sections, giving this hypothetical company a score or more than 800, it is unlikely it could become a winner with scores in the results categories so low.

The scores, individual categories, and total are mainly used in the early stages of the awards process. High-scoring applications are selected for the consensus review stage. High-scoring applications after consensus scoring are selected for site visits. After site visits, scores are *not* recalculated. The actual findings of the site visit teams are submitted to the judges, and the judges get further information from the site visit team leader or members. At this stage of judging, scores have become much less important and are rarely used. The site visit teams concentrate their activity on finding the evidence to support claims in the applications, verifying results, and examining supporting documents. These visits focus very much on results, not just approach or deployment. The focus is on whether the company's approach is working and is working across the company and across all functions. Examiners verify data, interview employees, and review actual operations and facilities.

During the site visit, examiners look for measurements of both internal and external quality. They look for measures of suppliers' quality levels. They interview employees and ascertain the results of the training, teamwork, and quality improvement processes. They look at customer satisfaction data, competitive evaluations, and benchmarks. They look for evidence of actual, sustained improvement and world-class performance.

Administration of the MBNQA

The MBNQA is administered through a complex set of processes under the management of the United States Department of Commerce, Technology Administration, NIST. Administration for the award is provided by the American Society for Quality Control (ASQC). Most of the actual work of reviewing and scoring applications, site visits, judging, and developing the management processes is done by several hundred volunteers from U.S. companies, universities, and government offices and consultants.

The Board of Overseers

The Board of Overseers is a small group of people who have established preeminence in quality management. For example, the recent chair of the Board of Overseers has been Robert W. Galvin, the chairman of the Executive Committee of Motorola. Motorola was one of the first winning companies of the MBNQA. Armand V. Feigenbaum, William A. J. Golomski, and Juran have all served as members of the Board of Overseers.

The overseers are concerned mostly with questions of process. They ensure that proper processes for managing the MBNQA are in place, are working, and are continuously improved. They review recommendations by the judges as to process improvements, but the overseers are not involved in the actual evaluation and judging of the applicants.

Issues of concern for the overseers include number of awards, award categories, changes to the Act, and technology sharing and transfer based on lessons learned.

The Board of Examiners

The Board of Examiners consists of more than 200 people selected according to expertise, experience, and peer recognition. They do not represent companies or organizations, but serve as volunteers for the common good. All members of the Board of Examiners receive three days of rigorous training using case studies, scoring exercises, and team building sessions. They become a powerful network for quality improvement throughout the United States.

The Board of Examiners consists of three distinct groups: judges, senior examiners, and examiners. There are nine judges. The judges oversee the entire process of administering the award, help select examiners, review the scored applications, select the organizations to receive site visits, and review the results of the site visits. They then decide which, if any, organizations to recommend for the MBNQA.

The final decision for the awards is made by the Secretary of Commerce after further background evaluations of the recommended organizations. These further evaluations are intended solely to determine if an organization is facing environmental charges, Justice Department action, or other problems. If these concerns are substantial, the secretary may remove the organization from the recommended list. The secretary may not add any organization to the list and has no other influence on the awards process.

The judges are involved in oversight at every stage of the MBNQA process, but only get involved in the review of actual applicants after many hours of work by the examiners. These evaluations, screenings, and site visits provide the foundation on which the award process is built.

There are approximately 20 to 30 senior examiners, and they play a crucial role. They are selected for their experience and expertise; many have been examiners for several years or have been directly involved in winning organizations' quality management. They score applications and manage the consensus review process.

There are almost 200 examiners who score all the applications, perform site visits with the senior examiners, and provide input each year on how to improve the application guidelines, the scoring process, and the entire awards process.

The Awards Process

The MBNQA process follows several carefully defined steps. The first is the annual improvement of the criteria, the guidelines, and the entire awards process. The next step is the completion of the eligibility determination form by the company.

Applicants must have their eligibility approved prior to applying for the award. The next step is the completion and filing of the applications by the individual companies or organizations. The award applications then go through four stages of review.

1. Independent review by at least five members of the Board of Examiners

2. Consensus review and evaluation for applications that score well in Stage 1

3. Site visits to applicants that score well in Stage 2

4. Judge's review and recommendations

The scoring system used by the Board of Examiners is described in the application guidelines. It is based on three evaluation dimensions: (1) approach, (2) deployment, and (3) results. All examination items require applicants to furnish information relating to one or more of these dimensions. The scoring guidelines are presented in Figure 3.4.

Each year, after the recommendations for the winning companies are forwarded to the Secretary of Commerce, the judges review the entire MBNQA process. Feedback is solicited from all members of the Board of Examiners, applying companies, ASQC, the staff of the National Quality Award Office, and other interested parties. The suggestions for improvement are carefully considered, and each year a number of changes are made to the award criteria, the application guidelines, and the award process. This constant improvement is one of the greatest strengths of the MBNQA.

In 1992, Reimann said the award criteria

> reflect much learning from past years. Improvements are numerous. We have given even greater emphasis than before to productivity, speed, waste reduction, and to long-term outlook. An expanded introduction to the criteria goes further than ever before in explaining the Award's values, characteristics, and goals. Each year the award program – all aspects of it, I might add – undergo rigorous review by our Board of Overseers. Our Board of Overseers looks to our panel of judges for advice in developing its recommendation.[6]

The Winning Companies – Lessons Learned

In the first seven years of the MBNQA, 22 organizations have received the award. The recognition these companies have received from the president of the United States, from the press, and from senior executives throughout the country has far exceeded

SCORE	APPROACH/DEPLOYMENT
0%	• Anecdotal information; no system evident in information presented
10% to 30%	• Beginning of a systematic approach to addressing the primary purposes of the item • Significant gaps still exist in deployment that would inhibit progress in achieving the major purposes of the item • Early stages of a transition from reacting to problems to preventing problems
40% to 60%	• A sound, systematic approach responsive to the primary purposes of the item • A fact-based improvement process in place in key areas addressed by the item • No major gaps in deployment, through some areas may be in early stages of deployment • Approach places more emphasis on problem prevention than on reaction to problems
70% to 90%	• A sound systematic approach responsive to the overall purposes of the item • A fact-based improvement process is a key management tool; clear evidence of refinement and improved integration as a result of improvement cycles and analysis • Approach is well-deployed; with no significant gaps, although refinement, deployment, and integration may vary among work units or system activities
100%	• A sound systematic approach, fully responsive to all the requirements of the item • Approach is fully deployed without weaknesses or gaps in any areas • Very strong refinement and integration – backed by excellent analysis

Figure 3.4. MBNQA scoring guidelines.

SCORE	RESULTS
0%	• No data reported or anecdotal data only • Data not responsive to major requirements of the item
10% to 30%	• Early stages of developing trend data • Some improvement trend data or early good performance reported • Data are not reported for many to most areas of importance to the item requirements and to the company's key performance-related business factors
40% to 60%	• Improvement or good performance trends reported in key areas of importance to the item requirements and to the company's key performance-related business factors • Some trends and/or current performance can be evaluated against relevant comparisons, benchmarks, or levels • No significant adverse trends or poor current performance in key areas of importance to the item requirements and to the company's key performance-related business factors
70% to 90%	• A good to excellent improvement trend in most key areas of importance to the item requirements and to the company's key performance-related business factors or sustained good to excellent performance in those areas • Many to most trends and current performance can be evaluated against relevant comparison, benchmarks, or levels • Current performance is good to excellent in most areas of importance to the item requirements and to the company's key performance-related business factors
100%	• Excellent improvement trends in most to all key areas of importance to the item requirement and to the company's key performance-related business factors or sustained excellent performance in those areas • Most to all trends and current performance can be evaluated against relevant comparisons, benchmarks, or levels • Current performance is excellent in most areas of importance to the item requirements and to the company's key performance-related business factors • Strong evidence of industry and benchmark leadership demonstrated

Source: National Institute of Standards and Technology, 1993 Awards Criteria, Malcolm Baldrige National Quality Award, Gaithersburg, Md., 1992.

Figure 3.4. *(continued).*

even the most optimistic hopes of the award's creators. The 22 winners have given more than 10,000 presentations about what they accomplished and how they achieved their results. The staff of the MBNQA Office, members of the Board of Examiners, and members of the Board of Overseers have also given more than 10,000 presentations about the award criteria, the award process, and the results of the winners.

Although the number of applications guidelines requested has been in the hundreds of thousands, the number of companies applying has been more modest, basically limited to those that felt they were far enough along their quality journey to benefit from a formal application and, of course, to those eligible to apply. The number of applicants in the three categories of manufacturing, service, and small businesses are given in Figure 3.5.

The winning companies for the first seven years of the MBNQA are given in Figure 3.6. Although the concentration of winners has been either manufacturing companies or divisions of manufacturing companies, in recent years a number of service companies has also won. The breadth of industries represented by both winners and applicants has been remarkable.

In their report on the status of TQM for JUSE, Juran and Godfrey point out several lessons from the MBNQA process.[7] The first is the stunning results achieved by the winning companies. The second is how these results have been achieved across the companies – in every function, in every work process, in every location. The third is that these results are being achieved across industries, service and manufacturing, and in small and large businesses.

Stunning Results

The results reported by the winners have been stunning. The most visible feature of those results is their stunning magnitude. Most companies in the past set modest annual goals of a few percent. The achieved gains of the MBNQA winners have been far greater. They have shared numerous case examples of gains in a few years that show

- The time to provide customer service has been reduced by an order of magnitude.

- Defect levels have been reduced by 10 times or even 100 times.

- Productivity has been doubled.

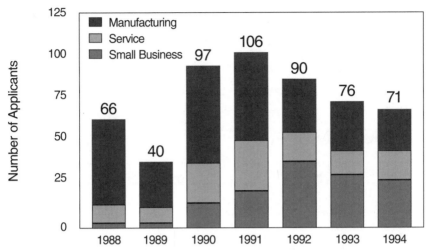

Source: *National Institute of Standards and Technology, Malcolm Baldrige National Quality Award, Gaithersburg, Md., 1993.*

Figure 3.5. The number of MBNQA applicants from 1988 to 1994.

- Costs have been cut by 50 percent.
- Returns from customers have been reduced to zero.
- All shipments have been made on time for 42 months.
- Product quality has been improved from 99 percent to 99.7 percent in two years.
- Reliability has been improved by an order of magnitude in five years.
- Product costs have been cut by 30 percent to 40 percent in two years.
- Manufacturing hours have been reduced by 58 percent in three years.
- Design cycle times have been reduced by 51 percent in three years.

Company	Category	Year
AT&T Consumer Communications Service	Service	1994
GTE Directories	Service	1994
Wainwright Industries	Small Business	1994
Eastman Chemical Company	Large Manufacturing	1993
Ames Rubber Corporation	Small Business	1993
AT&T Network Systems Group Transmission Systems Business Unit	Manufacturing	1992
AT&T Universal Card Services	Service	1992
Granite Rock Company	Small Business	1992
Texas Instruments Defensive Systems & Electronics Group	Manufacturing	1992
The Ritz-Carlton Hotel Company	Service	1992
Marlow Industries	Small Business	1991
Solectron Corporation	Manufacturing	1991
Zytec Corporation	Manufacturing	1991
Cadillac Motor Car Company	Manufacturing	1990
Federal Express Corporation	Service	1990
IBM Rochester	Manufacturing	1990
Wallace Co., Inc.	Small Business	1990
Milliken & Company	Manufacturing	1989
Xerox Corporation Business Products & Systems	Manufacturing	1989
Globe Metallurgical, Inc.	Small Business	1988
Motorola, Inc.	Manufacturing	1988
Westinghouse Electric Corporation Commercial Nuclear Fuel Division	Manufacturing	1988

Source: National Institute of Standards and Technology, Malcolm Baldrige National Quality Award, Gaithersburg, Md., 1994.

Figure 3.6.   MBNQA winners from 1988 to 1994.

During the presentations of the winning organizations, many executives from the leading U.S. companies learned for the first time how great the opportunities for improvement are in the average organization. They heard from Motorola how it had reduced defective solder connections from about 4000 ppm to between 1 ppm and 5 ppm. They heard how, in 40 months, Motorola's Communication Sector had reduced overall defects by 86 percent. Motorola had reduced warranty claims by 54 percent. In some specific parts of its operations, the results were even more stunning. Motorola achieved a 30 to 1 reduction in factory cycle time and a 10 to 1 improvement in field reliability. By the end of 1988, Motorola was calculating that it was spending $250 million less in cost of poor quality than in 1986.[8]

Despite Motorola's impressive results by the time it had won the MBNQA, it set even more ambitious goals for the next few years. It wanted another tenfold improvement by 1989 and to be at six sigma (approximately 3.4 ppm) quality levels by 1992. By the end of 1992 it estimated it had made, on average, a 150 times improvement over the levels it had attained when it won the award.

One of small businesses to win the MBNQA is Globe Metallurgical, a major producer of silicon metal and ferrosilicon products. From 1985 to 1988 it increased its market share of ductile iron products from less than 5 percent to greater than 50 percent. It has reduced its complaints from customers by a factor of 11 and the returns from customers to zero percent. Globe has quality improvement teams at all levels. Employee teams produce more than one implemented idea per employee per week. More than 70 percent of these ideas are implemented within 24 hours. Globe's product quality has become so legendary that many customers in the United States and Western Europe no longer give detailed specifications or requirements, they just ask for "Globe quality." While other American steel companies are struggling to survive, Globe Metallurgical is running 24 hours a day, seven days a week.

Another MBNQA winner was the CNFD of Westinghouse Electric Corporation. This division currently has more than 40 percent of the U.S. market for fuel assemblies and about 20 percent of the world market. It improved reliability of the fuel assemblies by a factor of 10 from 1985 to 1988, to 99.995 percent reliability. Its goal for 1989 was another tenfold improvement or 99.9995 percent. Another measure of its success is 42 straight months of on-time deliveries. In new product development, it has reduced its cycle time of six or seven years to only two years for new fuel assemblies.

It achieved more than 98 percent on-time and error-free software by 1988. In reliability its results have been spectacular. It has had no failures attributed to corrosion in 30 million feet of fuel tube construction. It has had no failures attributed to weld faults in 7.3 million welds. It has had no failures attributed to fuel densification in 600 million pellets. By 1988 it was defining specific customer satisfaction measures for each different customer. For the CNFD, all of this paid off handsomely. Between 1985 and 1988 it cut its warrantee costs by more than half; it increased its orders by more than 50 percent.

Milliken and Company, a 1989 MBNQA winner, epitomizes the team approach to quality. Milliken has 47 manufacturing facilities in the United States and produces more than 48,000 different textile and chemical products. In 1988 Milliken had more than 1600 corrective action teams address specific manufacturing or other internal business challenges. About 200 supplier action teams worked to improve Milliken's relationships with its suppliers. In addition, nearly 500 customer action teams responded to the needs and aims of customers, including the development of new products. While enjoying its success, Milliken – like Motorola – continues to set ambitious new targets for quality improvement. Its new goals include a tenfold improvement in key customer-focused quality measures in the next four years.

Another 1989 MBNQA winner was the Xerox Corporation's Business Products and Systems. It employs 50,200 people at 83 U.S. locations. During 1988, 7000 quality improvement teams within Xerox were credited with savings of more than $100 million by reducing scrap, tightening production schedules, and devising other efficiency and quality-enhancing measures. Working closely with its 480 suppliers, Xerox has reduced the number of defective parts reaching the production line by 73 percent in the last five years. Planning new products and services at Xerox is based on detailed analyses of data organized in 375 information management systems. One hundred seventy-five of the systems are specific to planning, managing, and evaluating quality improvement. Each year Xerox's customer service measurement system tracks the behavior and preferences of 200,000 owners of Xerox equipment.

In the past five years Xerox has achieved a 78 percent decrease in the number of defects per 100 machines and decreased unscheduled maintenance by 40 percent. Like other MBNQA winners, Xerox is also setting ambitious goals for the future. Its targets for 1993 include a 50 percent reduction in unit manufacturing cost and a fourfold improvement in reliability.

Results Across the Spectrum

The second notable feature of the results of the award winners is that the improvements took place throughout the entire spectrum of company activities: customer satisfaction, field performance of products, quality of the manufacturing processes, suppliers' quality, timeliness of customer service, quality of business processes, and employee safety.

Many executives were surprised at how widely the winning companies were applying quality management. Many still had a "little q" version of quality in their minds: conformance to specifications and requirements and zero defects. The Baldrige Award winners broadened their horizons. The winners focused on customer complaints, inventory reductions, and delivery times.

Federal Express, was a 1990 Baldrige Award winner. CEO Fred Smith said: "If a customer has a complaint or problem, we'll get back to him or her that day with a response, even if it's to say we're still working on it. The sun doesn't go down without a customer getting a reply to a question."[9]

Westinghouse had, for more than 20 years, attacked its inventory problem without any significant effect. It had focused on symptoms, without going after the disease itself. In 1982 it started a more systematic attack by focusing on the root causes of excess inventories. As it improved its factory processes, corporate inventories, as a percentage of sales, began dropping steadily and permanently.

Average gross inventories fell from 24 percent of sales in 1982 to 16 percent in 1989. This differential – from 24 percent to 16 percent – represents about $800 million of cash Westinghouse didn't have to spend on inventories.[10]

It used to take Motorola three weeks to both process an order for certain customized pagers and have the manufactured product ready for shipment to the customer. Motorola reduced this time frame to two hours. It has invested in flexible manufacturing systems, has developed close relationships with an elite group of suppliers who provide nearly real-time delivery, and has made many improvements in the decision chain. A salesperson now plugs in the customer's order on a handheld computer, and the data are transmitted electronically to the shop floor, where an empowered employee takes the information and starts the manufacturing process within minutes.

Motorola has extended its aggressive goals to all areas in the company: purchase orders, recruiting letters, invoices, typed reports, and so on. In fact, Galvin reported at Juran Institute's IMPRO '89 Conference that among the most spectacular quality improvement gains within Motorola in 1989 were those in the Patent Department.[11]

Customer loyalty has become one of the focus points of many of the leading companies. Zytec Corporation, a 1991 winner, reported that, of 26 new-billed customers since 1984, 20 are still customers and it has become the *sole* supplier for 18 of these. Zytec's annual original equipment manufacturer revenue growth rate has averaged 74 percent since 1984 versus an industry average of 8 percent to 10 percent. Zytec's customers also come to Zytec for repairs. Since 1988 it has had an annual revenue growth rate in this line of business of 45 percent.

It has achieved these results by dramatically improving product quality and reducing warranty costs by 48 percent, repair turnaround time by 31 percent, and product costs by 30 percent to 40 percent. It has also used impressive improvements in cycle times to become more responsive to customer needs. In four years it reduced design cycle times by 51 percent, safety approval times by 50 percent, and printed circuit board layout times by 69 percent. It has also considerably improved the product in the eyes of the customer by increasing mean times to failure by 10 times in five years.[12]

Solectron Corporation, another 1991 Baldrige Award winner, worked on its credit approval process to improve customer satisfaction. Its old cumbersome process was bureaucratic and time-consuming. It now quickly classifies customers as "A-rated" and no approvals are needed. "B-rated" customers are approved or disapproved quickly using a scoring sheet, and "C-rated" customers are not pursued.

Solectron also reported improvements in pricing consistency, unscheduled inventory to production issues, inventory accuracy (a $250,000 savings itself), overtime, price quotation complexity, and price change updates.

For the MBNQA winners, quality improvement efforts have paid off handsomely. Xerox has regained four percentage points in worldwide market share. In small copiers, an extremely competitive market, Xerox rebounded from a low of 8.6 percent market share in 1979 to 15 percent in 1986.

In the highly competitive business of nuclear fuel, the Westinghouse CNFD now has more than 40 percent of the U.S. market and more than 20 percent of the world market. In specialty metals, a market that most U.S. companies had abandoned to foreign competitors, Globe Metallurgical is capturing new markets throughout the world.

In another very competitive market, textiles, Milliken is recording record profits. In electronics, probably the most publicized threatened industry in the United States, Motorola has become the world leader in cellular telephones. Its miniature electronic pagers are selling very well in the United States and in Japan.

Results Across Every Industry

One of the most obvious lessons from the MBNQA winners has been the wide applicability of quality management to all industries. In the first five years of the Baldrige Award, winners included high-tech manufacturing companies, a textile company, a division of an automobile manufacturer, a nuclear fuel supplier, an automotive parts supplier, a distributor of oil industry equipment, a rock supplier, a hotel, a credit card company, and an overnight mail service. The winning companies have ranged from those with only 160 employees to companies employing tens of thousands of employees.

Although the identities of applying companies are usually kept secret, the 22, site-visited companies in the 1988 and 1989 awards process were listed by the GAO. Twenty of these companies participated in a special GAO study. These companies included a telecommunications products company, a car company, two chemical companies, a telephone company, two insurance companies, a furniture systems division, a clothing mail order company, three computer companies, several other high-tech companies, and several automotive suppliers or divisions.

The results shared by these companies left no doubt of the applicability of quality management methods across all industries. In fact, Allan I. Mendelowitz, director of GAO's International Trade, Energy and Finance Issues Division, who headed the GAO study, stated that the results were so impressive that the GAO intended to establish a TQM process of its own.[13]

How the Results Have Been Achieved

The results achieved by these companies were not only impressive, but they were achieved in a variety of ways. However, there was no set formula all the winning or site-visited companies used. They did share some traits in common. All demonstrated extraordinary leadership, especially at the senior executive level. All had established outstanding information and analysis systems. Their quality measurement systems were thorough, understood throughout the company, and used widely for decision making. And each of these companies involved members of the company at every level, in every function, and in every activity.

The common themes from the winning companies that have impressed audiences of executives from throughout the U.S. include the following.[14]

- Aggressive quality goals
- Benchmarks from within industry and from outside industry leaders
- Response time drivers
- Proactive customer systems
- Quantitative orientation, heavy investment in measurement systems
- Major human resource investments, training at every level, continuously

The senior management leadership has stood out in all of these companies; none has made these gains with a small, isolated department running things. The companies have been strongly customer driven. They are aware of their marketplace, their competitors, the changing environment, and the changing customer needs.

Leadership

In each of the winners of the MBNQA the leadership of the senior executives has been outstanding. This leadership goes far beyond the items contained in Category 1, "Senior Executive Leadership."

The senior leaders in the winning organizations have seen clearly the connection between quality and productivity, between quality management and company success. Smith put it quite clearly in a letter to the members of Federal Express.

> Recently, we achieved our highest recorded daily service level, 99.7 percent. Just as significantly, on that same day we had our lowest cost per package *ever*. Imagine, the very best service and the lowest cost!
>
> Our challenge is to find ways to continue delivering the highest quality service at increasingly lower costs. I believe we can do this and all evidence proves it can be done using these quality management tools.[15]

Another lesson learned is how winning companies have incorporated quality management into their strategic planning processes. As Ronald D. Schmidt, chairman, president, and CEO of Zytec, states, "Our original business plan, generated in 1983, stated we would differentiate ourselves from our competitors by the quality of our products."[16]

Zytec uses a specific process to do this called "Management By Planning." This Management By Planning

> is a methodology to align every employee in the company to be pulling in the same direction. It allows Zytec to take a five-year (long-range strategic plan) and convert it into four one-year objectives, which then are deployed to every department in the company for them to establish supporting one-year objectives with monthly goals to measure their progress against. All four of our 1992 corporate objectives are quality related.[17]

John Marous, chairman of Westinghouse, after traveling to the White House to receive the first MBNQA from President Reagan on behalf of the Westinghouse CNFD became virtually a total quality fanatic.[18] He commissioned a Total Quality Fitness Review (Westinghouse's internal version of the MBNQA) of his own chairman's office. Marous personally participated in the review of his office and later publicized the improvement teams which were assembled to improve processes as recommended by the review.

Marous never discussed financial performance when he made site visits throughout the Westinghouse business units. "All he was interested in was how the total quality journey was going. He knew that if total quality was progressing, the numbers would take care of themselves."[19]

At Marlow Industries, a 1991 Baldrige Award winner, senior management, like in all the winning companies, has taken charge of the quality system. Raymond Marlow, chairman and CEO, states

> Our Quality System is managed through the TQM Council. The council meets once each week for one hour. Council members consist of the CEO (who is chairman) and our five senior executives as permanent members, seven other employees that rotate annually, and invited guests. All members have equal voting rights. The TQM Council is the operational structure of our company. The council oversees and mentors the quality system; establishes, reviews, and assesses quality processes; and identifies areas for improvement.[20]

A keystone of quality leadership is effective communication throughout the organization. At Marlow Industries this is done in a wide variety of ways. Daily information and results are shared, by managers personally in the organization, internal/external customer results are shared and other quality information is available on the local area network. The TQM Council meets weekly and decisions and results are shared. Monthly company meetings allow council members to share company performance figures and results of quality activities, and to recognize employees. Marlow Industries publishes its own quarterly newsletter and holds an annual Quality Commitment Day.

Measurement

The winning companies have all demonstrated an almost fanatical obsession with measurement. Data abounds throughout the companies and information is available to every employee. Xerox has more than 200 information systems related directly to quality of products, of services, of internal processes, and of customer satisfaction. Federal Express has continuously refined its measurement techniques over the years. It has continuously sought to measure quality from the customer's point of view. Its Service Quality Indicators have helped it make extraordinary improvements in actual service. It feels there is absolutely no question that it measures service with a more reliable and more timely process than any of its competitors.[21]

Moreover, Smith believes that the Service Quality Indicators have helped Federal Express change its thinking, from looking at service quality as a percentage of on-time deliveries, to looking at it as the actual number of service failures. Smith believes this change in thinking has made a profound difference in Federal Express' own perceptions of how it is performing, and as a result it has radically improved its delivery of service to the customer.

These companies all have strong information systems. The necessary data are available, sophisticated analysis systems are in place, and the information is used. They have an excellent basis for assessing their own quality status and comparing their results with others. The measurements are interlinked and support corporate goals. Quality management activities are clearly linked to corporate strategic plans.[22]

Involvement of All Members of the Company

The empowerment of employees in the winning companies has also been surprising to some executives more used to the control-and-command style of management. One of the key strategies at Zytec highlights this.

Our Vice President of Marketing and Sales "empowers" every employee to spend up to $1000 to solve a customer problem and to send the bill to him.[23]

These companies have developed active, working plans to involve every member of the company. Employee surveys are used routinely to gather information about how employees feel about their work, environment, and job satisfaction. Employees openly identify areas for improvement. The goal is to have every person in the organization fully trained and actively working on solving problems.

Marlow Industries, one of the small business winners of the MBNQA, does this in a very structured way. It uses three basic training classes to prepare every company member. The first concentrates on professional qualification. Employees are trained to do their jobs through action and tests. The second involves effectiveness team training where employees learn the quality improvement process. And, the third training area involves in-depth training in the quality tools used in quality improvement and problem solving.[24]

These companies continue to make major investments in human resource development. Training and education budgets are growing from 2 percent or 3 percent of total revenue to 5 percent or 6 percent. The involvement of the workforce continues to grow.

For example, training at one of the winning companies averaged 85 hours for every employee in 1991. By 1992 it planned to average 95 hours, 110 in 1993, 125 in 1994, and 150 by 1995. A smaller company was averaging only 72 hours per employee, but some of the larger ones were averaging more than 100 hours per production employee and more than 160 hours for R&D employees.

The Impact of the MBNQA

There are strong indications that the MBNQA has become a major force in the United States to improve quality. Hundreds of thousands of copies of the application guidelines have been circulated. Tens of thousands of presentations and visits have been made. Scores of books, articles, and interviews have been published.

In many ways these presentations by award winners at conferences, company open houses, and to company executive sessions were benchmarks. They enabled a wide audience of American executives and managers to see new ideas and approaches. They had a certain shock value as managers compared their own results and approaches with other companies. They helped all executives, managers, and workforce members to see the gaps in the performance and the improvement potentials.

During 1991, the U.S. GAO completed a study of MBNQA winners and site-visited companies. It studied carefully the relationship between quality management activity and success and profitability. This report, GAO Report 91-190, became GAO's all-time, best-selling report.

This study further fueled the interest in the MBNQA. Although the GAO found it difficult to measure financial success as a function of quality management efforts, it did find a clear relationship between these companies' quality improvement activities and their marketplace successes.

> The Baldrige Quality Award has played a key leadership role in revitalizing the commitment of corporate America to total quality management.[25]

> The Malcolm Baldrige National Quality Award has been a tremendous action item on the part of the Department of Commerce. It has made the nation aware that government and business must work together to improve quality awareness and implementation if we trade in the world marketplace.[26]

Sharing – How Companies Are Learning from Each Other

The Baldrige Award is a demanding competition, with every company subject to the same stringent tests. Points are awarded for originality, and there are only six possible winners a year. One would expect these rules to produce clannishness and secrecy, as each company pursues its own gains.

> In fact, the results have been the opposite: an outpouring of cooperative behavior and a level of corporate sharing seldom seen in this country.[27]

Senior executives, managers, union leaders, and average workforce members are absorbing talks, papers, and broadcasts about quality management. Xerox talks to more than 100,000 people a year, many of them customers and suppliers. One executive from Globe made 138 speeches in the year following its winning. David A. Garvin quotes one of the examiners, "We absolutely don't believe this would have happened without the Baldrige Award."[28]

Garvin claims the award has created a common vocabulary and philosophy which has created bridges across companies and industries. Senior executives, managers, researchers, and employees now view learning across the boundary lines of business as both possible and desirable. Garvin feels we are now breaking down the "not invented here syndrome," which was once a source of corporate uniqueness and pride. We are now finding an unabashed zeal for borrowing ideas and practices from others.

> In many ways, this spirit of cooperation is the legacy of the Baldrige Award. Winners are compelled by law to share their knowledge; that they have done so without suffering competitively has led other companies to follow suit. Benchmarking is by definition a cooperative activity, and it is an award requirement. Even warring factions of the quality movement have united under the Baldrige Award banner. To become more competitive, American companies have discovered cooperation.[29]

Using the MBNQA for Self-Assessment

The greatest impact the MBNQA has had, by far, is its use as a self-assessment tool by thousands of companies and organizations throughout the United States. In this way the original goals of the Public Law are being realized. Quality awareness has been promoted to an extent far greater than expected, companies have shared successful quality strategies, and several thousand companies have taken a hard look at how they compare with the best.

The number of application guidelines requested from NIST gives an indication of this interest. In the first five years of the MBNQA this interest has been remarkable. In 1988, "only" 12,000 application guidelines were requested. By 1991 this number had grown to 285,000. In 1992, 175,000 were requested (see Figure 3.7). The actual number distributed is really not known, since NIST gives open permission to make copies of these guidelines. For example, in 1991, one company reproduced 30,000 copies internally to pass out to every employee.

Although the number of applicants has also grown, this growth has been far more subdued and limited mostly to companies that thought their quality management process was at least mature enough to warrant a site visit. Many companies did not intend to apply at first; they were just using the MBNQA criteria for self-assessment.

Federal Express, a 1990 Baldrige Award winner, is among a number of companies which originally went through a Baldrige self-assessment with only the goal of self-improvement in mind. But by the time the company completed the process, senior managers reasoned that they might at least rate a site visit, and they opted to apply.[30]

The Xerox Business Products and Systems Division also went through an intensive review process of its own in 1989. It states that it found 513 "warts," its internal name for what MBNQA examiners would call "areas for improvement." Xerox used these warts to create action plans to prepare for the examiners' site visit and for their five-year strategic planning process. Senior managers took responsibility and accountability for these action plans.

Schmidt described in testimony before the U.S. House of Representatives Subcommittee on Technology and Competitiveness of the Committee on Science, Space, and Technology how this had helped his company in the following way.

> Our sixth strategy . . . was to use the Malcolm Baldrige National Quality Award criteria to further improve our company. We used the knowledge gained from (our first application) to make some process improvements and reapplied in 1991, and as they say, "the rest is history."[31]

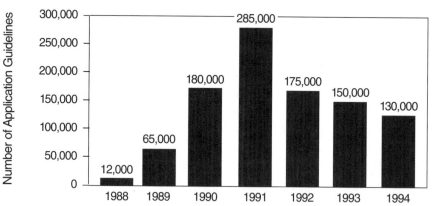

Source: National Institute of Standards and Technology, Malcolm Baldrige National Quality Award, Gaithersburg, Md., 1993.

Figure 3.7.   Number of MBNQA application guidelines requested from 1988 to 1994.

Smith, CEO of Federal Express, echoes Schmidt's remarks.

Applying for the Malcolm Baldrige National Quality Award is a major undertaking but it is an incredibly valuable experience. The process of completing the 75-page application provided us the opportunity to pinpoint our strengths and weaknesses in the area of quality improvement. Assessing areas that need improvement will help us develop future improvement plans – the biggest benefit gained by applying for the award.[32]

Westinghouse has created an intensive internal review process closely related to the MBNQA criteria. These reviews are conducted on a voluntary and confidential basis by a senior team from outside the unit being reviewed. At the end of the week-long review, the team gives the unit's management a score, a strengths and weakness analysis, and a set of recommended next steps for total quality improvement.

Westinghouse has had more than 450 of these reviews in the past 10 years. The reviews have provided "a marvelous indicator of progress, as well as an excellent tool for identifying specific improvement opportunities at each operating unit."[33]

Like a growing number of U.S. companies, Westinghouse conducts its own mini-Baldrige Award competition within the company. The ultimate recognition awards are the George Westinghouse Total Quality Awards. The prize for each is $200,000 – which can be used for any purpose except bottom-line enhancement. The internal winner receives strong consideration to be a candidate for the MBNQA the following year.[34]

The Westinghouse business units, like the divisions in many other companies, find the act of just applying for the internal award to be a worthwhile activity. One of its general managers, Jim Watson, states "We gained so much insight from researching the application requirements that, whether we get the award or not, we are winners!"[35]

These internal company assessments are becoming widespread in the United States. Chester Placek reports a number of examples in some of the leading companies.[36] Convex Computer Corporation makes the award a visible goal, sending a clear message to everyone that quality is the company's primary priority.

Dow Chemical has instituted an internal award patterned after both the MBNQA and the Deming Prize. AT&T has established the Chairman's Quality Awards as a process to encourage the company's business units and divisions to meet the MBNQA criteria. The Chairman's Quality Awards process is structured to encourage cooperation among AT&T's various businesses and for winning in the marketplace.

AT&T's first winners of the Chairman's Quality Awards in 1991 included its Universal Card Services, Information Management Services, Switching Systems and Transmissions. In 1992, two of these divisions, AT&T Universal Card Services and AT&T Transmissions Systems, won the MBNQA.

Like AT&T, IBM has made internal competition for its own award an integral part of its quality improvement strategy. At IBM Santa Teresa, the company's largest programming laboratory, an excellence council uses the Baldrige Award template as a road map and measurement vehicle. It consists of internal and external reviewers who examine, score, and report progress along the path to world-class TQM."[37]

Many companies that are widely recognized as leaders in their industries are also finding the MBNQA criteria useful. FMC Corporation, which was already achieving a better than 15 percent return on investment for five consecutive years and was the market leader, also has embarked on an intensive self-review program. Martin Dorio, director of quality and productivity, has described how this self-review helps FMC identify areas where it can further increase its quality improvement efforts. He states "The system aids in evaluating and setting priorities. It helps find gaps in company strategy and assists in the development of action plans."[38]

Impact on Government, Academia, and Other Awards

Government

In the Unites States the MBNQA has also helped stimulate much interest in TQM in the public sector. Although not eligible at the present time to compete for the award, government agencies at federal, state, and local levels are using the award criteria for self-assessments and for various award programs within the government. The U.S. Air Force recently announced its plans to use the MBNQA criteria as its definition of TQM and to perform self-assessments to these criteria widely.

As Reimann explained to Congress in 1992,

> The elevation of quality on the national agenda is reflected in growing awareness and action within the federal government. This includes not only the Department of Commerce and the Technology Administration but also many other parts of the government. During the past year we have held fruitful discussions with

representatives of the Federal Quality Institute, many units of the Defense Department, NASA, the Department of Labor, the Department of Education, and the Council on Environmental Quality. In addition, we had the opportunity to brief key leaders from several government-oriented professional organizations at a meeting held at the National Academy of Public Administration.[39]

Universities, Technical Schools, and Community Colleges

Many universities have become interested in the MBNQA. Courses on TQM are being added to business schools, engineering schools, and even medical schools throughout the country. A number of professors and even one business school dean has served on the MBNQA Board of Examiners. Some schools have even begun to apply TQM principles as well as teach them.

One of the stimuli of this change has been the University TQM Challenge Grant started by five leading U.S. companies (Motorola, IBM, Xerox, Milliken, and Procter & Gamble). Seven universities have been selected for support by these companies under this grant. Both business schools and engineering schools in these universities are participating in learning from these companies, changing their curriculums, and applying TQM principles in the day-to-day operations of the universities.[40]

Many technical schools and community colleges have set up centers for training small businesses in TQM concepts, methods, and tools and are also beginning to apply these ideas to their own operations. Many companies have provided support to these schools with gifts of materials and loans of instructors. Companies are seeing clear personal benefits in supporting these efforts.

Recently, steps have been taken to create a national quality award for engineering schools using inputs from the MBNQA office. The Rochester Institute of Technology (RIT) and *USA TODAY* have implemented a Quality Cup Award for individuals and teams. The first awards were presented in 1992.[41]

During his testimony to congress in 1992, Reimann, stated how rapid this interest in quality management has been in universities: "This past year, for example, there has been a tremendous upsurge in business school and, I might add, other academic interest."[42]

He went on to explain how in 1991 the American Assembly for Collegiate Schools of Business devoted its entire annual meeting to managing for quality and how he feels that the changes taking place in business schools are dramatic. This is especially remarkable in the United States since as recently as 1989 "business schools were perhaps seen as 'part of the problem.' This picture has changed dramatically in the ensuing three years."[43]

The introduction of quality management concepts is also happening in the secondary school systems in the United States. As Reimann reports,

> Many of the grassroots efforts around the United States are drawing in education institutions at all levels. The American Association of School Administrators (AASA), a professional association of school superintendents, reports that interest in quality has stepped up greatly.[44]

Figure 3.8 shows how the MBNQA relates to education.

State Award Programs

More than half of the individual states now have or are in the process of implementing their own version of the MBNQA. These award programs, for the most part, are closely based on the MBNQA criteria and process. Many use the same training materials for their examiners. The differences appear in the breadth of the award programs. The states are not limiting their awards only to for-profit companies, and many are including government agencies and local governments in the eligibility lists.

The New York State Quality Award, the Excelsior Award, is an excellent example of one of the working state award programs. The governor of New York feels that the issue is standard of living which is a function of the productivity growth rate. And the sources of competitive advantage for a state are natural resources, capital, technology, and skills. The seven criteria of the MBNQA address all these issues.

The state quality awards give smaller companies, the non-*FORTUNE* 1000 companies, a chance to compete and receive valuable feedback. New York, like many other states, has a number of problems. The Excelsior Award is intended to promote quality awareness and practices in businesses, educational institutions, and government agencies; recognize quality achievements; publicize quality strategies and programs; stress the importance of labor-management cooperation; and acknowledge the interrelationship of the public, private, and education sectors.

The first winners of the Excelsior Award were: Albany International, Press Fabrics Division; Motorola, Elma Plant; New York State Police; and Kenmore – Town of Tonawanda Union Free School District.

BALDRIGE AWARD KEY CONCEPTS The Concept	APPLICATION IN EDUCATION The Need
1. The customer defines quality. Customers are both the ultimate consumer and fellow employees.	Educator with students, parents, and the community all work to meet the ultimate customer's needs – the student.
2. Senior leadership defines and introduces values throughout the company.	Federal, state, and local policy makers provide consistent leadership, sense of purpose, support.
3. The company assures there are well-designed/executed systems and processes.	Organization and management systems must change to implement new policies and values. Systemic organizational changes occur at the state, district, and school levels.
4. Continuous improvement is a critical part of the management process.	Evaluation, including quantitative measures, is used continuously by those within the system to assess programs and make improvements.
5. Goals and operating plans are developed consistent with values and purpose.	National, state, and district educational goals are established and operational plans put into place to support these goals. Practices, funding, and staff development flow from the goals.
6. Shortened response times are critical to the company's ability to meet ever-changing customer needs.	Decision-making rests as close to the student as possible. Support is given to school-based management and shared decision-making.
7. Operations and decisions are based on facts and data.	Operations and decisions are based on community needs, demographics data, and information on school and student performance.
8. All employees must be well trained and involved in quality activities.	Staff development is provided as a critical component of systemic change to improve student performance.
9. Success hinges on quality and error prevention.	The education system is designed so that all children learn (for example, provision of early childhood and at-risk programs, varied learning environments, and pedagogical approaches).
10. Companies communicate quality requirements to suppliers.	Educators reach out to parents and human service agencies serving students and families.

Source: Raymond Wachniak, Total Quality Curriculum, *Vocational Industrial Clubs of America, Leesburg, Va., 1993.*

Figure 3.8. Applying Baldrige Award concepts to education.

A Look Back

In the past 12 years there has been an increasing global emphasis on quality management. In global competitive markets, quality has become the most important single factor for success.

Reimann, in his testimony to the U.S. Congress, stated this clearly. "There is now far clearer perception that quality is central to company competitiveness and to national competitiveness."[45] Reimann pointed out the changes happening from his vantage point.

> I would say that in 1991 – and continuing now into 1992 – we are seeing stronger signs of action, and stronger signs of quality concepts taking root. At the same time, we are seeing continued growth in quality awareness in all sectors of the U.S. economy.[46]

In the United States, the president and the Secretary of Commerce have given their personal support and attention to quality, thus elevating quality on the national agenda. Their efforts have helped the American public become aware that quality is a main component in national competitiveness.

Most American executives have found the comprehensive definitions of quality used by the winning companies to be real eye-openers. These executives are beginning to make changes in their own organizations. They no longer relegate quality management to a small specialized department. They, instead, are taking on the task of quality leadership personally. They are creating strategic quality plans, they are developing quality information and analysis systems, they are improving human resource management, they are investing in training and education for all members of the organization, and they are focusing on improving customer satisfaction. Even more importantly, these same leaders are learning the strategies and means used by the winning companies to change entire corporate cultures and obtain world-class competitive positions. Managers all over the United States have stated how amazed they were upon learning the actions and results of the winning companies. A common remark is, "I did not have the slightest idea what total quality management really is."

Recently, Garvin prepared an in-depth summary of the MBNQA for the *Harvard Business Review*. In this summary Garvin reports "an outpouring of cooperative behavior and a level of corporate sharing seldom seen in this country."[47] He goes on to state,

In just four years, the Malcolm Baldrige National Quality Award has become the most important catalyst for transforming American business. More than any other initiative, public or private, it has reshaped managers' thinking and behavior. The Baldrige Award not only codifies the principles of quality management in clear and accessible language. It also goes further: it provides companies with a comprehensive framework for assessing their progress toward the new paradigm of management and such commonly acknowledged goals as customer satisfaction and increased employee involvement.[48]

All the evidence that we have points to the enormous impact of the MBNQA on U.S. competitiveness. In just seven years this award has sparked a quality revolution that is changing in fundamental ways how American companies are managed. We are just beginning to glean a fraction of the lessons to be learned from the winning companies. But the leaders of companies throughout the United States are already building new strategies based on these early lessons to change entire corporate cultures and attain world-class competitive positions.

There is now almost a national mania for self-assessment to the MBNQA criteria, for benchmarking against competitors, and for benchmarking best-in-class practices. U.S. companies are now on fast learning curves.

The true strength of the MBNQA is that it is a living award. Each year the examination process has gotten better. Each year the criteria and areas to be addressed have been improved. Each year the applications have been better, the results even more stunning, the winning companies even more deserving. Each year we have learned more, each year we have more to learn.

Notes

1. Malcolm Baldrige National Quality Award, National Institute of Standards and Technology, Route 270 and Quince Orchard Rd., Administration Bldg., Rm. A537, Gaithersburg, MD 20899, U.S.A., (301) 975-2036, fax (301) 948-3716.

2. National Institute of Standards and Technology, *Hard Copy. 1992 Basic Set*, MBNQA presentation slides, Gaithersburg, Md., 1992.

3. Tito Conti, "A Critical Review of the Current Approach to Quality Awards," Proceedings of the EOQ Conference, Brussels, Belgium, June 1992, pp. 130-139.

4. National Institute of Standards and Technology, *1993 Application Forms and Instructions,* Malcolm Baldrige National Quality Award, Gaithersburg, Md., 1992.

5. Ibid.

6. Curt W. Reimann, testimony on the Malcolm Baldrige National Quality Award before the Subcommittee on Technology and Competitiveness of the Committee on Science, Space, and Technology, U.S. House of Representatives, U.S. Government Printing Office, February 5,1992.

7. J. M. Juran and A. Blanton Godfrey, "Total Quality Management (TQM) – Status in the U.S.," proceedings on the Senior Management Conference on TQC, JUSE, Tokyo, November 1990.

8. Ibid.

9. Federal Express Corporation, *Information Book,* description of Federal Express' 1990 Malcolm Baldrige National Quality Award-winning application process and employee information, 1990.

10. Jack Fooks, "Focus on the Malcolm Baldrige National Quality Award (Part III), Total Quality 'Lessons Learned,'" *The Quality Management Forum,* Vol. 17, No. 3, Fall 1991, pp. 1-3, 15.

11. Robert W. Galvin, "Role of the Chief Executive in Quality Management," dinner speech at IMPRO 89, Juran Institute, Atlanta, Ga., November 1989.

12. Zytec Corporation, *Zytec Application Summary,* summary of Zytec's 1992 MBNQA winning application, 1991.

13. Marion Mills Steeples, *The Corporate Guide to the Malcolm Baldrige National Quality Award* (Homewood, Ill.: Business One Irwin, 1992), p. 38.

14. Curt W. Reimann, "The First Five Years of the Malcolm Baldrige National Quality Award," proceedings of IMPRO 92, Juran Institute, Wilton, Conn., November 11-13, 1992.

15. Fred Smith, "Our License to Practice," reprinted in Federal Express Corporation's *Information Book,* December 5, 1990.

16. Ronald D. Schmidt, testimony on the Malcolm Baldrige National Quality Award before the Subcommittee on Technology and Competitiveness of the Committee on Science, Space, and Technology, U.S. House of Representatives, U.S. Government Printing Office, February 5, 1992.

17. Ibid.

18. Fooks, "Focus on the Malcolm Baldrige National Quality Award."

19. Ibid.

20. Raymond Marlow, testimony on the Malcolm Baldrige National Quality Award before the Subcommittee on Technology and Competitiveness of the Committee on Science, Space, and Technology, U.S. House of Representatives, U.S. Government Printing Office, February 5, 1992.

21. Smith, "Our License to Practice."

22. Reimann, "The First Five Years of the Malcolm Baldrige National Quality Award."

23. Schmidt, testimony on the Malcolm Baldrige National Quality Award.

24. Marlow, testimony on the Malcolm Baldrige National Quality Award.

25. National Institute of Standards and Technology, *Hard Copy. 1992 Basic Set.*

26. Marlow, testimony on the Malcolm Baldrige National Quality Award.

27. David A. Garvin, "How the Baldrige Award Really Works," *Harvard Business Review* (November-December 1991): pp. 80-93.

28. Ibid.

29. Ibid.

30. Steeples, *The Corporate Guide to the Malcolm Baldrige National Quality Award.*

31. Schmidt, testimony on the Malcolm Baldrige National Quality Award.

32. Smith, "Our License to Practice."

33. Fooks, "Focus on the Malcolm Baldrige National Quality Award."

34. Ibid.

35. Ibid.

36. Chester Placek, "Baldrige Award as a Quality Model," *Quality* (February 1992): pp. 17-20.

37. Ibid.

38. Ibid.

39. Reimann, testimony on the Malcolm Baldrige National Quality Award.

40. Galvin, "Role of the Chief Executive in Quality Management."

41. Placek, "Baldrige Award as a Quality Model."

42. Reimann, testimony on the Malcolm Baldrige National Quality Award.

43. Ibid.

44. Ibid.

45. Ibid.

46. Ibid.

47. Garvin, "How the Baldrige Award Really Works."

48. Ibid.

Chapter 4:

Federal Quality Institute Awards for Federal Government Groups and Employees

WILLIAM A. J. GOLOMSKI

The Presidential Award for Quality was initiated about the same time as the MBNQA, but it is not as well-known as the MBNQA. It is presented by the president of the United States and is administered by the Federal Quality Institute in Washington, a governmental organization which helps federal units to improve using TQM. Two other categories of awards are presented at the same award ceremony. The awards were initiated in 1987 and presented for the first time in 1988, except for the Presidential Quality Award, which was first presented in 1989. These three awards are eligible only to federal government units and employees.

President's Council on Management Improvement Award for Management Excellence

The President's Council on Management Improvement (PCMI) Award for Management Excellence recognizes federal employee, group, and agency excellence. This is in support of the President's Management Improvement Program and overall improvement in the quality of federal operations and services to the public. Those eligible nominate themselves. The maximum number of awards that may be given each year is two.

The number of PCMI Awards given each year in the past is

- 1992 – 9
- 1991 – 11
- 1990 – 9
- 1989 – 14
- 1988 – 15

The awards are presented by a cabinet official.

Here are a few examples of programs than won in 1992.

Model Workplace Program
Executive Staff and the Office of the Assistant Secretary for
   Administration and Management
Department of Labor
Washington, DC

The Model Workplace Program has been established to explore and evaluate new employment practices to better meet current and future workforce needs. The program has resulted in better mission performance, improved recruitment and training, reductions in on-the-job injury and illness rates with "lifetime" savings of $20 million in worker compensation, reduction by nearly 10 percent in employee use of sick leave adding 50 years in productive time, and the use of the program as a breakthrough technique in labor-management relations.

Notice Improvement Quality Action Team
Office of Disability and International Operations
Social Security Administration
Department of Health and Human Services
Baltimore, MD

The Notice Improvement Quality Action Team streamlined and automated the process in which beneficiaries are notified of the person or agency that will be responsible for managing or directing the management of their benefits. The original six-step process requiring 70 minutes and four people was reduced to four steps, 16 minutes, and two people. Case costs were reduced about 65 percent – from $24.83 to $6.64 or $3.7 million annually.

Jesse C. Smiley
Granite Goose Project Office
U.S. Army Corps of Engineers
Department of Defense
Pomeroy, WA

Jesse C. Smiley suggested an improved method for emergency shutdown of water flow at hydroelectric dams. Although initially intended as a way to improve the survival rate of migrating fish, the suggestion resulted in the cancellation of dam construction that saved $8.9 million and another $19.6 million in cost avoidance due to the cancellation of two other dam sites scheduled for reconstruction. The fish survival rate was also increased 10 percent to 15 percent with almost a million additional juvenile fish reaching the Snake River.

QIP Award

The annual QIP Award, administered by the Federal Quality Institute, was created to

- Recognize federal organizations that have successfully adopted TQM principles and thereby improved the quality, timeliness, and efficiency of their services or products

- Provide models for the rest of government, demonstrating that a commitment to quality leads to better services and products and more satisfied customers

An applicant organization must meet several conditions. It must

- Be a part of the federal government, staffed by general employees

- Have no fewer than 100 full-time employees

- Be autonomous, with its own defined mission

- Provide products or services to customers outside the organization's own agency [except for Department of Defense (DOD) organizations, whose primary customers are often other military organizations]

- Provide mission-related services or products

Each cabinet department and executive agency may submit a maximum number of applications for the QIP Award, according to its size.

- Up to 20,000 employees – two applications.

- Up to 50,000 employees – three applications.

- Up to 95,000 employees – four applications.

- More than 95,000 employees – five applications.

- The DOD may submit a maximum of five applications for *each* service (Army, Navy, and Air Force) and a total of four applications for the other defense agencies.

In addition, each agency may submit one application for an administrative or support organization. The application must cover an entire function, not just a branch or division. For example, an application could be submitted for an entire finance department, but not for a payroll office that might be part of the department.

These limits are set to encourage agencies to conduct internal reviews and select the best nomination. The award process should not be used as a substitute for agency organization assessments. QIP Award winners are ineligible to compete for four subsequent years. For example, winners of the 1989 QIP Award will not be eligible to reapply until 1994. An application that does not satisfy these requirements will not be evaluated. The applicant will be notified.

Timetable

Applications available	May 1992
Applications due	August 31, 1992
Application review	September 15-17, 22-24, 1992
Site visits	September 29 - November 20, 1992
Final judging	December 15-16, 1992
Awards presented	May 1993

Application

A standardized nomination form must be completed giving demographics and the name of organizations. Then a report following the award criteria must be attached. The report must not exceed 35 pages. Seven pages of illustrative attachments (charts, graphs, quality vision statement, and so on) may be added to the 35 pages. Pages in excess of these limits will not be examined. The application must be typed or printed in no smaller than 10-pitch or 12-point font. Each subelement of the criteria (for example, la., lb., and so on) must be separately labeled and addressed in the application.

An additional one- to three-page summary should be included that states the mission of the applicant organization, describes its products and services, identifies the customers of its products and services, and provides a brief history of when and why TQM was implemented. This summary is not counted as part of the 35-page limit for the written report. It is particularly important to describe the mission, products, and services so that individuals unfamiliar with the organization can fully understand its business.

The application must be able to stand on its own. Answers for each subelement should be fully responsive and assume no prior knowledge of the organization. Responses should be concise and quantitative when possible. Statements should be supported by facts and information. Assertions unsupported by plausible data, information, or facts will receive no credit during application evaluation. Care should be taken to fully define acronyms and terminology specific to the business of the applicant organization. If acronyms are used, define them and and provide a glossary (glossary pages will not be counted in the previously stated page limits).

The examiners and judges are customers in the application review process. Feedback is gathered directly from them. Applicants, the suppliers in the application review process, should strive to satisfy the customers; they have expectations of an application package that is easy to evaluate. Doing so will optimize the evaluation and selection processes, as well as enable the examiners to produce a thorough feedback package that accurately reflects the applicant's status in relation to the criteria.

Submission of Applications

Agencies should designate a centralized coordinator responsible for submitting applications. The coordinator must ensure that the quota for submissions is observed and that all applicants meet the eligibility requirements. Agency submissions should be sent to the Federal Quality Institute as a package.

To apply, mail or deliver 10 copies of the complete application package (including nomination form) to

> *Quality Improvement Prototype Award*
> *Federal Quality Institute*
> *National Building Museum, Room 333*
> *401 F Street, N.W.*
> *Washington, DC 20001*

Deadlines

The Federal Quality Institute must receive applications by August 31, 1992. Applications received after this date are ineligible for the 1993 award.

Evaluation Process

An evaluation system based on points is used by the examiners recruited from public and private sector organizations. A maximum of 10 finalists receive site visits by examiners. Judges will select a maximum of six prototypes. A feedback report is sent to each eligible applicant. All finalist and prototype organizations will receive plaques, but prototype organizations will receive additional recognition. They will be promoted as showcase organizations throughout the year.

Award Criteria

The criteria are the basis for applying for the award and providing feedback to applicants. The criteria define a quality system – the key elements of a quality improvement effort and the relative importance and interrelationship of these elements. The criteria embody certain fundamental concepts of TQM.

- Quality is defined by the customer.

- The organization is driven by continuous improvement.

- The focus is on prevention of errors rather than detection.

- Everyone participates in quality improvement.

- Senior management creates quality values and builds the values into the way the organization operates.

- Employees are valued and recognized for their involvement and accomplishments.

The criteria elements are

- Top management leadership and support (20 points)

- Strategic quality planning (15 points)

- Customer focus (35 points)

- Training (10 points) and recognition (5 points)

- Employee empowerment and teamwork (20 points)

- Measurement and analysis (15 points)

- Quality assurance (30 points)

- Quality and productivity improvement results (50 points)

The point distribution indicates the relative importance of each element of the criteria in an integrated TQM system.

The number of QIP Awards given in the past were

- 1992 – 5

- 1991 – 2

- 1990 – 3

- 1989 – 6

- 1988 – 4

Two QIP Overviews

1992: Public Services and Administration
Patent and Trademark Office
Department of Commerce
2121 Crystal Dr., Suite 908
Arlington, VA 22202

Public Services and Administration (PSA) is responsible for public services; preliminary and post-examination processing of patent applications; and traditional administrative services related to procurement, mail processing, space, telecommunications, and other support services.

In 1989, senior managers – faced with increasing workloads, costly rework, and high employee turnover – began to dispense with traditional management and to change their entire organizational culture. They worked with their 750 employees to launch quality management under the banner of TEAMWORK. It stands for *Together – Employees and Managers Work*. PSA has been named a 1992 prototype based on its many achievements.

- Average days to mail patent application filing receipts has been reduced from 37 days in fiscal year 1988 to 18 days in fiscal year 1992. Quality was improved, reducing rework by more than 50 percent and public requests for corrections by 81 percent.

- Processing time to record patent ownership was reduced from more than 100 days processing time to 20 days, with a 50 percent reduction in errors.

- Productivity in the mail room increased by 63 percent over fiscal year 1988 and saves $1 million annually.

- Customer perception of the procurement staff's responsiveness to customer needs improved from 72 percent in fiscal year 1989 to 93 percent in fiscal year 1991.

- In just one year, employee turnover decreased from 24 percent to 10 percent.

1988: Naval Aviation Depot
Department of the Navy
Cherry Point, NC

The Naval Aviation Depot, Cherry Point, has brought improved quality to the repair of sophisticated aircraft, engines, and associated aeronautical components. Improvement was achieved through management leadership, better communication among managers and employees, gainsharing incentives, involvement of employees, and intensive training in SPC and other analytical skills for assessing improvement in work processes. Quality improvement resulted in a decrease in repair cost for final product outputs – components, engines, and aircraft. For example, cost trends indicate that the average cost for the repair of an engine component dropped as much as $7000 per unit from 1985 to 1987. The cost of an engine repair was reduced as much as $3000 between January 1987 and September 1987. Standard repair and maintenance on one type of aircraft dropped as much as $55,000 from January 1986 to September 1987. Through the gainsharing incentive, employees were able to share in the savings produced by these cost reductions. Employees received 46 percent of the more than $1.8 million saved in the first quarter of 1988.

Aftermath

Interest about the winners continues to be high. U.S. Marine Colonel J. B. Gartman, who was the commanding officer, retired and is now a consultant to other governmental groups. Several of the others in the staff continue their efforts in TQM and want to win the Presidential Award for Quality. John S. W. Fargher Jr., Department Head of Management Controls, was the founder of the successful national conference on TQM in government in 1992. It is sponsored by IIE.

Presidential Award for Quality

This is the highest award for federal government quality. It, the other awards mentioned, and the MBNQA were initiated in 1987 during the final term of office of President Reagan. The annual Presidential Award for Quality, administered by the Federal Quality Institute, was created to

- Recognize organizations that have implemented TQM in an exemplary manner, resulting in high quality products and services, and the effective use of taxpayer dollars

- Promote TQM awareness and implementation throughout the federal government

An applicant organization must meet several conditions. It must

- Be a part of the federal government, staffed by federal employees.

- Have no fewer than 500 full-time employees.

- Be autonomous, with its own defined mission.

- Provide products or services to the American public (with the exception of DOD organizations, whose primary customers are frequently other military organizations). An administrative or support organization is not eligible.

Each cabinet department and executive agency may submit a maximum number of applications for the Presidential Award according to its size.

- Up to 50,000 employees – two applications.

- Up to 95,000 employees – three applications.

- More than 95,000 employees – four applications.

- The DOD may submit a maximum of three applications for *each* service (Army, Navy, and Air Force) and a *total* of three applications for the other defense agencies.

These limits are set to encourage agencies to conduct internal reviews and select the best nominations. The award process should not be used as a substitute for agency organization assessments.

Winners of the Presidential Award, including organizational subcomponents, may not reapply for four subsequent years. For example, the winner of the 1989 Presidential Award – the Naval Air Systems Command – cannot reapply until 1994. NAVAIR's depots and other subcomponents are similarly ineligible. An application that does not satisfy these requirements will not be evaluated. The applicant will be notified.

Timetable

Applications available	May 1992
Notification of intent to apply	July 24, 1992
Application due	September 25, 1992
Application review	October 14-15, 1992
Site visits	October 26 - December 18, 1992
Final judging	January 13-14, 1993
Awards presented	May 1993

Application

A standardized nomination form must be completed giving demographics and the name of organizations. Then a report following the award criteria must be attached. An organization with fewer than 20,000 employees should prepare an application of no more than 40 pages. Ten pages of illustrative attachments (charts, graphs, quality vision statement, and so on) may be added. An additional one- to three-page summary should be included that states the mission of the applicant organization, describes its products and services, identifies the customers of its products and services, and provides a brief history of when and why TQM was implemented. Attach an organization chart. Pages in excess of these limits will not be examined

An organization with 20,000 or more employees should prepare an application of no more than 50 pages. Ten pages of illustrative attachments (charts, graphs, quality vision statement, and so on) may be added. In addition, a one- to five-page summary should be included that describes the organization's mission, products and services, customers, and a brief history of when and why TQM was implemented. Attach an organization chart. Pages in excess of these limits will not be examined.

Applications must be typed or printed in no smaller than 10-pitch or 12-point font. Each subelement (for example, la., lb., and so on) of the criteria must be separately labeled and addressed in the applications.

The application must be able to stand on its own. It is particularly important to describe the mission, products, and services so that individuals unfamiliar with the organization can fully understand its business. Descriptions for each subelement should be fully responsive and assume no prior knowledge of the organization. Responses should be concise and quantitative when possible. Statements should be supported by facts and information. Assertions unsupported by plausible data, information, or facts will receive no credit during application evaluation. Care should be taken to fully define terminology specific to the business of the organization. If acronyms are used, define them and provide a glossary (glossary pages will not be counted in the previously stated page limits).

The examiners and judges are customers in the application review process. Feedback is gathered directly from them. Applicants, the suppliers in the application review process, should strive to satisfy the customers; they have expectations of an applications package that is easy to evaluate. Doing so will optimize the evaluation and selection processes, as well as enable the examiners to produce a thorough feedback package that accurately reflects the applicant's status in relation to the criteria.

Submission of Applications

Agencies should designate a centralized coordinator responsible for submitting applications. The coordinator must ensure that the quota for submissions outlined here is observed and that all applicants meet the eligibility requirements. Agency submissions should be sent to the Federal Quality Institute as a package.

To apply, mail or deliver 10 copies of the complete application package (including nomination form) to

Presidential Award for Quality
Federal Quality Institute
National Building Museum, Room 333
401 F Street, N.W.
Washington, DC 20001

Deadlines

Agency coordinators must notify the Federal Quality Institute which organizations will apply for the Presidential Award on or before July 24, 1992. A review will be made of each applicant to verify its eligibility. The Federal Quality Institute must receive applications by September 25, 1992. Applications received after this date are ineligible for the 1993 award.

Evaluation Process

An evaluation system based on points is used by the examiners recruited from public and private sector organizations. As for the prototypes, scoring guidelines are based on the approach and implementation (see Figure 4.1).

A maximum of five finalists will be selected based on the application scores. Examiners will conduct site visits to supplement and validate information contained in the application. Finalist organizations may be asked to bear site visit costs for up to three examiners. A panel of judges, also selected from the public and private sector, select a maximum of two winners from the finalists. The judgment is based on evaluations of the written applications and the results of site visits. A feedback report is given to each applicant.

Approach	Implementation
100% • World class; sound, systematic, effective; • TQM-based; continuously evaluated, refined, improved • Complete integration • Innovative	**100%** • Fully in all areas, functions • Ingrained in organization culture
80% • Well-developed, tested; TQM-based • Excellent integration across functions	**80%** • In most areas, functions • Evident in culture of most groups
60% • Well-planned, documented; sound, systematic,TQM-based; all aspects addressed • Good integration across functions	**60%** • In many areas, functions • Evident in culture of many groups
40% • Beginning of sound, systematic, TQM-based effort; not all aspects addressed	**40%** • Begun in some areas, functions • Evident in culture of some groups
20% • Beginning of TQM awareness • No integration across functions	**20%** • Beginning in some areas, functions • Not part of culture

Figure 4.1. Scoring guidelines for the Presidential Award.

The award is a custom-designed cut crystal trophy. It is presented at the National Conference on Federal Quality held in Washington annually.

Award Criteria

The criteria are the basis for applying for the award and providing feedback to applicants. The criteria define a quality system – the key elements of a quality improvement effort and the relative importance and interrelationship of these elements. The criteria embody certain fundamental concepts of TQM.

- Quality is defined by the customer.
- The organization is driven by continuous improvement.
- The focus is on prevention of errors rather than detection.
- Everyone participates in quality improvement.
- Senior management creates quality values and builds the values into the way the organization operates.
- Employees are valued and recognized for their involvement and accomplishments.

The criteria elements are

- Top management leadership and support (20 points)
- Strategic quality planning (15 points)
- Customer focus (35 points)
- Training (10 points) and recognition (5 points)
- Employee empowerment and teamwork (20 points)
- Measurement and analysis (15 points)
- Quality assurance (30 points)
- Quality and productivity improvement results (50 points)

The point distribution indicates the relative importance of each element of the criteria in an integrated TQM system.

The number of Presidential Awards given in the past were

- 1992 – 1
- 1991 – 1
- 1990 – 0
- 1989 – 1

Executive Summary from 1992 Winner

The winner of the Presidential Award in 1992 was the Ogden, Utah, Internal Revenue Center. In 1985, Ogden Service Center was caught in a crisis, along with the entire Internal Revenue Service (IRS). This crisis was due to the nation's drive for production at any cost. Ogden Service Center recognized the need to refocus its priorities and channel the energy of its people in a new direction. It initiated TQM principles throughout its organization. Because its employees soon realized it takes both managers and employees working together to achieve quality, the IRS changed the definition from TQM to Total Quality Organization.

Incorporating quality into an organization the size of the Service Center was a complex task. During the 1991 filing season, 6300 federal employees processed more than 26 million tax returns, collected more than $100 billion, and processed refunds for taxpayers totalling $9 billion. Many other important functions are performed in the center, including the correction of errors, answering taxpayer correspondence, and ensuring compliance to tax regulations.

In order to begin transforming its immense organization into a Total Quality Organization, it has implemented a series of interlocking driving principles designed to generate continuous improvement. These principles are

1. **Structure.** The structure includes councils, a value system, quality improvement teams, task teams, and ways to measure quality progress. To date, quality improvement teams have saved $3,730,959 and quality initiatives have saved $7,571,789. This gives the Service Center a total cost benefit of more than $11 million. Its partnership with the National Treasury Employees Union strengthens the overall structure of its quality improvement process.

2. **Commitment.** Commitment from all levels – top management, the union, and employees – accelerates the progress of the quality movement.

3. **Education.** Emphasizing quality in its training effectively communicates the quality vision to employees.

4. **Customer Focus.** Ogden Service Center reexamined the word *customer.* To improve customer service, it literally started from the inside out by identifying its internal and external customers and their needs.

5. **Involvement.** Involvement expands the quality improvement process. The Service Center finds that involving its employees in the process gives them the buy-in necessary to generate teamwork and pride of ownership. Every employee is part of the team that evolves into a Total Quality Organization.

6. **Recognition.** Recognition is an important principle used to generate enthusiasm in the workforce. Odgen has many ways to recognize its employees as they make quality happen through their day-to-day accomplishments. The celebration of success is limited only by the imagination and creative abilities of its employees.

The force Odgen generates through its driving principles lets it face the challenges of change. Odgen employees realize the quality process is not an easy one, but has many rewards. As long as they remain focused on their quest for excellence, they will provide quality service to their customers.

Aftermath

The 1989 and 1991 winners of the Presidential Award are continuing their quality improvement efforts. They are continuing to make improvements. They have not been distracted in other directions.

Chapter 5:

NASA Excellence Award for Quality and Productivity

RAYMOND WACHNIAK

Aim for excellence and reward those who persevere. These are the tenets for the NASA Excellence Award for Quality and Productivity.

In an effort to help the U.S. aerospace community maintain its competitive edge in the world market, NASA initiated a Productivity Improvement and Quality Enhancement (PIQE) program. In 1985, the PIQE initiative was expanded to further institutionalize quality practices with the creation of the NASA Excellence Award. The purpose in presenting the award is to

- Increase public awareness of the importance of quality and productivity to the nation's aerospace industry and the nation's leadership position overall

- Encourage domestic business to continuously pursue efforts that enhance quality and increase productivity which will strengthen the nation's competitiveness

- Provide a forum for sharing the successful techniques and strategies used by applicants with other American organizations

Renaming the Award – George M. Low Trophy

In 1990, the NASA Excellence Award for Quality and Productivity was renamed for George M. Low, a former NASA deputy administrator whose contributions to the U.S. space program exemplified a quality philosophy that was far ahead of its time.

The current NASA approach to quality management reflects and builds on the precepts conceived by this distinguished scientist and educator more than 30 years ago. With the George M. Low Trophy, NASA continues the vision of excellence by recognizing those organizations that demonstrate a singular commitment to quality, encourage continuous improvement, and demonstrate sustained excellence, a focus on customer satisfaction, and outstanding achievement in a TQM environment.

In recognition of Low's efforts the text on the trophy reads

> This Trophy is awarded in memory of George M. Low, who greatly contributed to the early NASA Space Program during 27 years of Government Service.

The medallion, which is in the shape of an Apollo Command Module, has alloyed in it a portion of an artifact flown to the moon and back on Apollo 11 – the first manned lunar landing mission on July 16-24, 1969. The Low Trophy is awarded to companies, both large and small, whose programs meet or exceed outstanding achievement in a TQM environment.

Candidate Eligibility

All NASA contractors, subcontractors, and suppliers are eligible irrespective of size or nature of their products/services within certain limitations. A candidate is defined as a facility/organization having a NASA contract/subcontract and meets the criteria as a large business or a small business. Some criteria apply to both large and small businesses, namely: the organization must be within the United States; the organization should function as a self-sustained profit center with a majority of its resources in one location; and the candidate should be considered as the facility/organization with the NASA contract or subcontract, rather than the entire corporation.

The following criteria differentiate large from small candidates.

- A large business must aggregate sales to NASA or a prime contractor, for the preceding three years that exceed $1 million, with at least $250,000 of sales in each of the three years. And, there should be a minimum of 50 full-time employees (or 100,000 employee hours) engaged in NASA work.

- A small business must aggregate sales to NASA or a prime contractor for the preceding three years that exceed $250 in each of the three years. And, there should be a minimum of 25 full-time employees with at least one-third of the employees engaged in NASA work.

Small divisions of large corporations are presumed to receive corporate support and/or resources and thereby qualify as a large business.

Selection Process

The total selection process, from submission of a candidate nomination letter through to the presentation of the award, requires about 13 months. The milestone schedule for 1992 is shown in Figure 5.1. Major steps in the process are outlined here.

Nomination Letter

The purpose of the nomination letter is to determine if a candidate is qualified to continue the evaluation process. Each candidate is required to submit information to permit verification by the NASA Evaluation Committee. Upon acceptance of the nomination letter those selected candidates are notified of approval, and preparation of the application report begins.

October 1991
Award application guidelines available.

December 2, 1991
Candidates submit nomination letters to ASQC with brief statements of eligibility compliance.

January 2, 1992
Evaluation Committee completes review of candidate. This includes review by field installations(s) and prime contractor(s) if the candidate is a subcontractor. Candidate is notified of committee's decision.

March 2, 1992
Successful applicant submits application report (35-page maximum) to ASQC.

May 1, 1992
Evaluation Committee reviews application report to select finalists based on whether candidate's organization commitment and accomplishments meet the award standards.

June-August 1992
On-site visits to finalists' organizations.

August 1992
Evaluation Committee meets to review results of on-site validation visits and prepare findings for review by the NASA TQM Steering Committee.

October 1992
Selection of annual award recipient(s) made by NASA administrator based on recommendations of the TQM Steering Committee.

November 1992
Finalists recognized at reception at Ninth Annual NASA/Contractors Conference. NASA administrator announces award recipient(s).

November-December 1992
Presentation of award by NASA administrator in special ceremony held at recipients' location.

Figure 5.1.   1992 milestone schedule for candidates applying for the
George M. Low Trophy.

Application Report

Candidates that have been verified as eligible applicants are required to submit sufficient information so that a complete and thorough evaluation can be made by the Evaluation Committee. The application report must follow the sequence of the criteria and the subelements for either a large or small business, as applicable.

The purpose of having separate criteria for small businesses is to acknowledge the difference in documentation and availability of resources between large and small businesses. The best organizations, irrespective of size, will have processes that address all major criteria areas described in the NASA guideline document.

Evaluation Criteria

Major evaluation criteria deal with performance achievements and process performance achievements for the past three years.

1.0 Performance Achievements

1.1 Customer Satisfaction – emphasis in this element is on measurable and verifiable satisfaction of NASA and/or prime contractor requirements for overall organizational performance.

1.2 Quality – emphasis in this element is on qualitative, quantitative, and substantiated accomplishments in both the design and delivery of quality products and services with an emphasis on continual improvement.

1.3 Productivity – the focus in this section is on demonstrated quantifiable increases in output per unit of invested resource.

2.0 Process Performance Achievements

2.1 Commitment and Communication – the emphasis in this section is on demonstrated leadership in establishing a quality culture. The necessary process changes to empower employees at all levels and eliminate organizational barriers to continuous improvement must be documented.

2.2 Human Resource Activities – the focus here is on the quantitative evaluation of the program and activities that are necessary to recognize the value of people to an organization.

A complete outline of the criteria and the value of points assigned each subelement is shown in Figure 5.2. Each subelement is scored based on the guidelines shown in Table 5.1.

Site Visits

Based on the results of the application report review by the Evaluation Committee, applicants that have demonstrated excellent performance in quality and productivity are selected for recognition as finalists in the award process and receive a site visit.

The data gathered at the site by the Validation Team are reviewed by the Evaluation Committee. No material can be forwarded for consideration after the validation visit is completed. The Evaluation Committee prepares and presents a findings report to the NASA TQM Steering Committee.

Evaluation Criteria Elements	Total Points
1.0 PERFORMANCE ACHIEVEMENTS	**600**
1.1 Customer Satisfaction	
1.1.1 Contract performance	120
1.1.2 Schedule	50
1.1.3 Cost	50
1.2 Quality	
1.2.1 Quality Assurance (hardware/software/service)	120
1.2.2 Vendor quality assurance and involvement	50
1.2.3 External communication	40
1.2.4 Problem prevention and resolution	40
1.3 Productivity	
1.3.1 Software utilization	40
1.3.2 Process improvement and equipment modernization	30
1.3.3 Resource conservation	30
1.4.4 Effective use of human resources	30
2.0 PROCESS PERFORMANCE ACHIEVEMENTS	**400**
2.1 Commitment and Communication	
2.1.1 Top management commitment/involvement	100
2.1.2 Goals, planning, and measurement	80
2.1.3 Internal communication	40
2.2 Human Resource Activities	
2.2.1 Training	50
2.2.2 Workforce involvement	50
2.2.3 Awards and recognition	40
2.2.4 Health and safety	40
TOTAL POINTS	**1000**

Figure 5.2.   Summary of evaluation criteria for the George M. Low Trophy.

Table 5.1. Scoring guidelines for the George M. Low Trophy.

Each criteria element is scored based on these guidelines. The determining percentage is the applied to the available points.

Percentage	Description	How long in place	Deployment	Performance	Resources	Planning
91-100%	Excellent	3+ years	91-100%	Sustained high performance with constant improvement.	Resources dedicated to activities are commensurate with need and effectiveness.	All activities are incorporated in master plan to meet specific needs with provisions for feedback and modification.
81-90%	Very Good	3 years	81-90%	Starts moderately and improves to high performance.	Most resources are adequate but some are excessive, inadequate, or ineffective.	Most activities are included as part of overall plan with some exceptions. Feedback and program modification provisions are not completely implemented.
71-80%	Good	2-3 years	61-80%	Gradual continual improvement.	Most resources are adequate but many are excessive, inadequate, or ineffective.	Most activities are incorporated in overall plan but many activities have no coordinator.
61-70%	Average	2 years	41-60%	Starts low to moderate and improves slightly.	Many areas have adequate resources but some are neglected entirely or poorly utilized.	Individual plans govern most activities but lack coordination. Feedback provisions are incomplete.
51-60%	Fair	1-2 years	21-40%	Starts low and improves to moderate.	Resources are allocated sparingly without proper regard for need or appropriateness.	Planning is sporadic although targeted for completion. No provisions for feedback or modification.
< 50%	Poor	<1 year	0-20%	Starts and stays low.	Most programs and activities are poorly supported.	Planning efforts are barely initiated.

Award Recipient Selection

There is no limit to the number of finalists that can be selected as award recipients. Selection of the annual award recipient(s) is made by the administrator on the recommendation of the NASA TQM Steering Committee based on its review of the findings report from the Evaluation Committee. All finalists selected as award recipients are announced during the Annual NASA/Contractors Conference. (All decisions of the administrator are final. Award recipients are eligible to apply for another award four years after receiving the award.)

Recognition

The receipt of the prestigious George M. Low Trophy carries with it the recognition by NASA that the award recipient has demonstrated sustained excellence and outstanding achievements in quality and productivity in the aerospace industry. The award signifies the recipients(s) provide products/services at such a high quality level that they set new levels of customer expectation.

Benefits of Participation

The benefits that applicants derive from participating in the award program are shared by an even broader business community. The award provides a unique learning environment for all organizations that seek to improve their products and services. Notably, benefits are reaped from the leading experience of applicants that assess themselves against the criteria, develop strategies to improve, and then willingly share not only the successful strategies, but also the lessons learned from approaches that failed. The insight provided from both perspectives has proven to be invaluable to quality improvement.

Previous recipients report the effort was well rewarded. A sample of comments from applicants who have been through the self-analysis process is shown in Figure 5.3.

The award program is managed by the NASA Quality and Productivity Improvement Programs Division and is jointly administrated by NASA and ASQC.

The most valuable part of the process is self-analysis.
These past award recipients testify to the contribution of their organizations.

"Applying for and receiving the George M. Trophy benchmarked our quality and productivity performance while stimulating our people to even greater expectations."

Thomas S. Marotta
President and CEO
Marotta Scientific Controls
Montville, NJ
(1990 recipient – Small Business)

"The George M. Low Trophy process provided a target for our improvement initiatives. It helped us sharpen our focus on areas where improvement opportunities existed."

R. G. Minor
President
Space Systems Division
Rockwell International Corporation
Downey, CA
(1990 recipient – Large Business)

"The NASA Excellence Award is a tribute to our people and their unrelenting commitment to improving quality, productivity, and customer satisfaction."

Robert B. Young, Jr.
President and CEO
Lockheed Engineering and Sciences Company
Houston, TX
(1989 recipient)

"The NASA Excellence Award provided us the impetus to define our internal quality and productivity benchmarks from which we can implement a continuous improvement program."

Robert D. Pastor
President
Rocketdyne Division
Rockwell International Corporation
Canoga Park, CA
(1988 recipient)

Figure 5.3.   Benefits of participating in the George M. Low Trophy award process.

Chapter 6:

IIE Award for Excellence in Productivity Improvement

KENNETH E. CASE

The IIE Award for Excellence in Productivity Improvement recognizes companies and organizations worldwide which, through diligent and innovative means, have accomplished significant, measurable, and observable achievements that increased productivity, eliminated human drudgery, and/or improved quality.[1]

The award, first given in 1980 by IIE, was established to recognize and reward the application of excellent efforts resulting in long-term continuous improvement and to disseminate information on the accomplishments awarded. It recognizes innovative organizations that make the best use of existing philosophies to develop their own productivity and quality management approach. The award's ultimate goal is the improvement of private and public organizational competitiveness.

Up to four awards may be given each year in the following categories.

- Large manufacturing company/organization
- Small manufacturing company/organization
- Large service company/organization
- Small service company/organization

Fewer awards will be presented if no company/organization meets high standards for the award criteria within a specific award category.

Eligibility

This annual award program is open to all companies and organizations, including subsidiaries or divisions within a company or organization, throughout the world.

A small company/organization must meet two of the following three criteria.

1. Less than 1000 employees
2. Less than $20 million in annual sales
3. Less than $30 million in total assets

If a subsidiary or division of a company/organization is nominated that meets the small firm definition, the operating manager must have full authority over operations, profit and loss, funds allocation, and primarily serve either the public or business other than the parent company/organization to assure independent operations.

A large company/organization must meet two of the following three criteria.

1. More than 1000 employees
2. More than $20 million in annual sales
3. More than $30 million in total assets.

A manufacturing company/organization must make a product suitable for use from raw materials and/or premanufactured components by hand or machine according to specifications with a division of labor. Examples include the following industries: aerospace and defense, automotive, chemical, computers, consumer products, food processing, electronics, furniture, packaging, and steel.

A service company/organization must provide a service that contributes to the well-being of consumers or business customers. Examples include the following industries: banking, insurance, utilities, retail, transportation, and distribution.

Criteria

There are three statements of general criteria.

1. The achievement shall be an outstanding accomplishment which was the result of
   - Disciplined methodology extending over a period of time (for example, the result was not a windfall or stroke of luck)
   - Technological development, invention, or application
   - Interdisciplinary approaches

2. The achievement resulted in
   - Productivity improvement far in excess of the national norm
   - Elimination of drudgery
   - The accomplishment of tasks or work not previously possible
   - Improved products or services
   - Improved quality

3. The achievement does not have to be the result of the application of industrial engineering techniques per se, but must be within the scope of the following definition of industrial engineering.

Industrial Engineering is concerned with the design, improvement, and installation of integrated systems of people, material, information, equipment, and energy. It draws upon specialized knowledge and skills in the mathematical, physical, and social sciences together with the principles and methods of engineering analysis and design to specify, predict, and evaluate the results to be obtained from such systems.

The award criteria are not meant to be prescriptive. They are guides to judge the applicant's success in matching both diligent and innovative efforts to each organization's unique culture and environment for continuous improvement.

It is assumed that the business environment is different for each organization. As such, a custom-tailored approach for productivity and quality improvement is essential to obtain full benefits and ensure a superior competitive edge. Each application is judged on its own merits and not penalized if all examples listed in the criteria are not in practice.

To allow for equitable evaluation of nominated companies/organizations for the IIE Award, regardless of their management philosophy, size, or type of industry, the award is judged by the seven criteria listed in Figure 6.1 (each criterion is equally weighted). Numerous specific examples for each of the seven criteria are given in the award application guidelines.

Application and Selection Process

The total selection process, from submission to award presentation, requires about six months. A complete application must be submitted by November 15. Evaluation and decision making will take place such that award winners will receive the presentation at the IIE Annual International Industrial Engineering Conference held in May of the following year.

To ensure the integrity of the award, to verify the accuracy of applications, and to provide useful feedback to companies/organizations submitting applications for the award, these procedures are followed.

1. Nominations must be approved by the local IIE chapter president, IIE division director, or other IIE officer as appropriate.

2. It is required that five individuals must be listed on the application who can verify the information contained within the application package.

1. Strategic and Business Planning – How the overall management philosophy demonstrates its commitment to planning for productivity and related quality improvement.
2. Productivity and Quality Leadership – The extent of personal involvement by top management in creating an environment for accelerating productivity and quality improvement efforts.
3. Productivity and Quality Measurement and Analysis – The primary analytical tools used to identify root causes of problems, pinpointing areas of concentration, implementation of improvements, and measurement of results.
4. Productivity and Quality Management – How the productivity and quality improvements are integrated into the management system.
5. Productivity and Quality Training and Continuing Education – The training and continuing education presently being conducted to promote productivity and quality awareness, employee skills improvement, and management competence.
6. Measurement of Results – Trends, as shown through the use of data, from the last five years (data must not be proprietary and must be independently verifiable.)
7. Innovation – Innovative efforts and breakthroughs attributed to improved competitiveness and elimination of human drudgery.

Figure 6.1. Summary of criteria for the IIE Award for Excellence in Productivity Improvement.

3. The two highest scoring applicants in each award category will be visited by the Special Productivity Projects Committee members or selected volunteers to review the finalist's nomination package, examine supporting data, and walk the floor to ensure the productivity and quality improvement philosophy is evident at the shop-floor level. The travel expenses of site visit teams will be the responsibility of the finalist being audited. Every effort will be made to organize site visit teams that will incur the lowest travel expense and avoid any conflicts of interest. (Total site visit expenses are typically less than $2000 per company including all travel and per diem.)

4. All finalist organizations visited by a site visit team will receive a written feedback report. The feedback report will be issued by the committee on strengths, weaknesses, and opportunities for improvement in the company's/organization's productivity and quality efforts.

Benefits of Participation

The rigor of the criteria and evaluation process ensures that winners improve productivity by improving quality. In fact, the criteria make sure productivity and quality are virtually inseparable. The criteria of planning, leadership, analysis, management, training/education, results measurement, and innovation provide excellent guidance. Therefore, participating firms, their employees, their customers and suppliers, and the public in general all gain from this award process.

Fargher, chairman of the IIE Special Productivity Projects Committee, says

> I would certainly recommend that any organization that has made substantial progress and/or has developed unique and innovative approaches to achieve productivity and quality improvements should most definitely apply. It should be noted that all organizations are eligible, be they nonprofit, academic institutions, government, or commercial.

Some recent comments from past winners appear in Figure 6.2. The list of all winners appears in Figure 6.3.

Note

1. This award is part of the Honors and Awards program of IIE. Copies of the complete criteria and the application form are available from IIE, 25 Technology Park, Atlanta, Norcross, GA 30092, U.S.A.

"During the last 10 years . . . employees have made substantial efforts to find better ways to do their jobs without sacrificing quality. These activities have generated total expense reductions of $650 million."

James F. Hoffmeister
Vice President, Administration
Anheuser-Busch
St. Louis, MO
(1985 winner)

"Improved productivity, improved quality of working conditions. . . . It's worth the effort."

Jesse J. Webb
President and CEO
Atlantic Steel Company
Atlanta, GA
(1985 winner)

"Our main accomplishment . . . enabled us to complete orders with lot sizes as small as one within 24 hours of receipt of the order."

Robert Jasenovec
Senior supervisor
Allen-Bradley
Milwaukee, WI
(1987 winner)

"Since winning the IIE Award . . . NAVAVNDEPOT is recognized as the leader and as a role model within government in productivity and quality improvement. Savings have approached 25 percent and are climbing."

John S. W. Fargher, Jr.
Naval Aviation Depot
Marine Corps Air Station
Cherry Point, NC
(1988 winner)

"Putting industrial engineering tools in the hands of our front-line supervisors through their use of personal computers . . . automation of processes were technologically and economically feasible . . . significant improvements in productivity."

Shahad Saeed
Director, Human Resources and Industrial Engineering
Mountain Fuel
Salt Lake City, UT
(1990 winner)

"Do it! . . . the assessment process is invaluable!"

David M. Autrey
Director, Total Quality
Texas Instruments
Johnson City, TN
(1991 winner)

Figure 6.2. Comments on the IIE Award for Excellence in Productivity Improvement from past winners.

1980	San Antonio Air Logistics Center
1981	Robert E. Fowler (General Electric)
1982	A. Joseph Tulkoff (Lockheed—Georgia)
1983	Black and Decker Corporate Management
1984	Boeing Commercial Airplane Company—Renton Division Diablo Systems
1985	Anheuser-Busch, Inc. Atlantic Steel Company
1986	Chrysler Corporation
1987	Allen-Bradley Company Baltimore Gas & Electric Company
1988	Naval Aviation Depot—Cherry Point Dayton Power and Light
1989	Norfolk Naval Shipyard
1990	Ford Motor Company Mountain Fuel Supply Company
1991	Texas Instruments—Johnson City, TN

Figure 6.3. Past winners of the IIE Award for Excellence in Productivity Improvement.

Chapter 7:

RIT/*USA TODAY*
Quality Cup Award for
Individuals and Teams

WILLIAM A. J. GOLOMSKI

RIT and *USA TODAY,* a national daily newspaper, started the RIT/*USA TODAY* Quality Cup Award in 1991 (see Figure 7.1). The first awards were given on April 10, 1992 at the Arlington, Virginia, headquarters of *USA TODAY.* Up to five national winners in total may win the award. They may be teams or individuals. This award is different from other quality awards that recognize entire companies, divisions, or departments.

The examination process has three parts. The first is an initial screening for completeness and likelihood of contention. The second is a detailed review of the application by a panel of judges, with a telephone consensus on who should get a site visit. The third level is that of a judge making a site visit. Meetings are held with the team leader, team members, and people from the training staff and senior management. The judges' report is accepted by the heads of the two sponsoring organizations (see Figure 7.2).

Prior to the award ceremony, *USA TODAY* reporters visit the sites of the winning teams and the finalists. Both groups, plus various managers and executives of the companies, are present on the day of the presentation. Winners get a handcrafted silver cup. Finalists receive plaques. More than 220 teams applied in 1992; there were five winners and 11 finalists.

The 1992 awards were presented by Thomas Curley, president and publisher, *USA TODAY,* and Richard N. Rosett, dean, College of Business, RIT.

Here are the comments of the person who introduced the winners.

1. First, from Advanced Circuits – Hopkins, Minnesota – I am happy to welcome Robert Zimmer, who represents a team of 10 people. Advanced Circuits is a manufacturer of printed circuit boards. The team that Zimmer represents developed a quality process that can produce film that requires no post-processing touch-up. According to Advanced Circuits information, no other circuit board manufacturer has achieved this goal.

2. Arnoff Moving and Storage Company – Poughkeepsie, New York – The finalist team of three people is represented by Michael Arnoff. In 1989, the Arnoff Company had a 3.7 percent error rate on a contract with IBM. Through the efforts of its team, this rate was reduced to zero by 1991. In its quest for quality, Arnoff has improved inventory control, reduced cycle time for processing, reduced error rate, improved availability, and reduced manpower requirements.

3. Association for the Blind and Visually Impaired – Rochester, New York – David Brown represents a team of 11 people. The nonprofit group operates a manufacturing facility employing 110 blind and visually impaired individuals. The facility manufactures more than 50 products. In 1990, the tubing department operated at a $28,000 loss. Through the quality activities of the team in 1991, the tubing department was able to eliminate overtime and to improve other key indicators. This resulted in a profit of $34,000.

4. Beth Israel Hospital – Boston, Massachusetts – Beth Israel Hospital is represented by Karen Cummings. This hospital's team of 21 people perfected a quality system that virtually eliminated the waiting time for patients needing diagnostic scans. Prior to the team analysis, a wait of five to 55 minutes had been the norm. With the new system, patient anxiety has been reduced, outpatients are given better care, and risks are reduced.

5. Federal Express Branch – Indianapolis, Indiana – A nominated team of eight people is represented by Donald Hardy. This Federal Express branch conducted a quality war on missorts, or stray packages. The quality process developed by the team resulted in the reduction of missorts providing an expense savings of more than $1 million for a six-month period.

6. Nyma – Greenbelt, Maryland – A team of 11 persons is represented by Azmat Ali. Nyma is a small business with 441 full-time employees providing management consulting and computer-based systems development expertise to government and commercial clients. The Nyma team developed a quality program for reducing errors in its software systems being delivered to the Federal Aviation Administration (FAA) on an IBM subcontract. In 1991, only six out of 130 system delivered had Nyma-induced errors. This resulted in Nyma being given 86 percent of available awards from IBM for quality software delivery. Nyma is targeting 100 percent of awards by 1993.

7. Shenandoah Industries – Greencastle, Indiana – Dennis Hamilton represents a team of seven people. The firm manufactures injected molded decorative automotive trim for major automotive manufacturers. The team developed and implemented a prelaunch planning system to deliver parts for Chrysler's highly successful minivan. With a two-year quality planning effort, the team was able to deliver an initial order of 20,000 pieces of product with a zero rejection rate. During the past 12 months the defect percentage has been five times lower than the customer's goals.

8. Shure Brothers – Evanston, Illinois – Bill Stotler will represent a team of 11 people from the firm's Texas plant. Shure Brothers produces microphones and electronic radio products. Through a quality process, the team was able to make some significant changes in the production of a microphone product that had been in the line for several decades. As a result of the team's effort, Shure was able to reduce inventory from 31,500 pieces to 2000 pieces, cycle time from 32 days to two days, and total labor from 56 people to 29 people.

9. Thompson Consumer Electronics – Bloomington, Indiana – Paul Dyer will represent a team of six people. Thompson has responsibility for final assembly of RCA and General Electric television sets and other products. It also functions as a central distribution center for other RCA and General Electric products. This team, called the Wild Five, reduced rejects for unseated television set anodes. The quality work of the team resulted in rejects being reduced from 0.28 percent to 0.04 percent. Savings to the firm amounted to $190,000 annually.

10. University of Michigan Hospitals – Ann Arbor, Michigan – A team of 12 persons will be represented by Mary Decker Staples. The University of Michigan Hospitals is part of the 886-bed University Medical Center. The team attacked the problems concerned with long waiting periods for hospital admissions. In 1987, a significant number of scheduled patients were waiting up to two hours for admission. Some were not admitted because a bed was not available. By 1991, the team had reduced the wait period to 18 minutes. It is also estimated that the budget was reduced by $260,000 as a result of the changes recommended by the group.

11. Weyerhaeuser Mortgage Company – Woodland Hills, California – A team of seven persons is represented by Tracy McKeon. The firm is a Federal Housing Administration (FHA)/Veterans Administration (VA) Conventional Lender. Through a quality improvement process, the team was able to reduce its processing of government claims from 120 days to 22 days. This speeded the firm's lending process and saved more than $100,000 a year in finance charges.

It is now my pleasure to present the five recipients of the silver sculpture cups.

1. The first cup for the government category goes to Ann Combs who represents a team of seven members at the Navel Air Station, Patuxent River, Maryland. This group examined travel voucher processing and, through the use of a quality system, was able to generate a savings in tax dollars of more than $42,000 a year. Employees using the system rated it with a score of 3.87 on a four-point scale.

2. The cup in the manufacturing category is awarded to U.S. Steel, Gary Works. Jeff Grunden represents a team of five hourly persons. The team improved the quality and service being provided to the firm's automotive customers. The group was able to help customers reduce coil damage, eliminate costly scrap problems, and reduce defects from 14 percent to zero defects.

3. The next cup for the nonprofit category is awarded to a team of nine represented by Mary Lucas Blunt from Sentara System in Norfolk, Virginia. Sentara is a network of healthcare providers. It gives quality care through four hospitals and other facilities. The Sentara Group, through its quality improvement team, implemented solutions to problems that decreased the turnaround time to physicians for radiology diagnostic test results. It reduced time from 73.5 hours to 13.8 hours, representing an 81 percent improvement in efficiency.

4. The service category cup this year is awarded to Federal Express at Memphis, represented by Melvin Washington who headed a team of 11 people. Washington's team, through quality approaches, improved the sorting system significantly. This improved customer service and generated a monthly savings of $70,000. In addition, the process has helped to motivate the entire Memphis hub to seek further improvements in the system.

5. The 1992 Quality Cup Award for small business goes to LS Electro-Galvanizing Company in Cleveland, Ohio. Donald Vernon represents a team of 13 people. The firm is in the business of zinc coating steel sheet for critical automotive and truck applications. Through quality integrated process management approaches, the team has been able to reduce customer claims costs from more than $8 a ton to $1.29 a ton, an 85 percent improvement. More importantly, this has met the industry world-class benchmark set by a Japanese firm.

Dear Executive:

RIT and *USA TODAY* are beginning their search for individuals and small teams who make significant contributions to the improvement of quality products and/or services in an organization.

Five national winners will receive the RIT/*USA TODAY* Quality Cup for Individuals and Teams. We invite you to nominate qualified individuals and teams on this form, which you may duplicate if necessary.

The Quality Cup competition presents a unique opportunity to bring together business and education to focus on the important issue of quality. The individuals and teams to be honored will exemplify the qualities RIT's College of Business emphasizes when educating its students and which guide *USA TODAY* in its reader/customer-driven approach to newspaper publication.

The RIT/*USA TODAY* Quality Cup, a national award, is different from other quality awards that recognize entire companies, divisions, or departments because it stresses the important fact that quality begins with the level of the individual.

The RIT/*USA TODAY* Quality Cup is a valuable sculpture – a solid silver cup, which sits atop a white marble cylinder with a solid silver disk at its base. It was created by Leonard Urso, an internationally recognized silver craftsman and associate professor in RIT's School for American Craftsmen. Five cups will be awarded April 10, 1992, at the ceremony at *USA TODAY* headquarters near Washington, D.C. Winners also will be recognized in special pages of *USA TODAY*.

Nominations must be received no later than December 15.

Thank you for your help. We look forward to receiving your nominations

Thomas Curley   Richard N. Rosett
President and Publisher   Dean, College of Business
USA TODAY   RIT

Figure 7.1.   Application packet for the RIT/*USA TODAY* Quality Cup Award for Individuals and Teams.

RIT/*USA TODAY* Quality Cup for Individuals and Teams

PURPOSE:

The best way for companies and other organizations to compete and thrive today is for them to adopt the principles of total quality management. Both RIT and *USA TODAY* support and endorse the quality movement in the United States. This award recognizes individuals and teams who make significant contributions to the improvement of quality products and/or services in an organization. Through this award they will become national role models to enhance the quality movement. The RIT/*USA TODAY* award, with its focus on individuals and teams, is different from other quality awards, which recognize entire companies, divisions or departments.

WHO IS ELIGIBLE:

Awards will be made in five organizational categories: service firms, manufacturing firms, not-for-profit organizations and institutions, governmental units and agencies of government, organizations with fewer than 500 employees in any of the prior categories. U.S. and foreign firms are eligible, as long as the employees nominated work in the United States. Individuals and teams – preferably comprised of 15 individuals or fewer – in the organizations are eligible. More than one nomination per organization can be submitted.

HOW TO NOMINATE:

The nominations should be made in recognition of exemplary customer service, an exceptionally valuable improvement in a system or process for achieving customer satisfaction, or for an exceptionally valuable improvement in a procedure for solving problems. This nomination form must be submitted for each team or individual. signed by the chief executive of the parent firm, non-profit executive director or senior governmental official. It must be accompanied by a $100 application fee (make checks payable to RIT College of Business) for each nomination. No one affiliated with RIT or Gannett Co., publisher of *USA TODAY,* may enter.

JUDGING:

Judging will be administered by RIT. Judges – quality experts selected by RIT—will personally visit with all finalists. Decisions of the judges are final.

DEADLINE:

Nominations must be postmarked by December 15, 1991. Please contact Carol Skalski (703/276-5890) if additional information is needed. When completed, send nominations to:
                                    Quality Cup
                                    c/o Carol Skalski
                                    USA TODAY
                                    1000 Wilson Blvd., 22nd floor
                                    Arlington, VA 22229

Figure 7.1. *(continued).*

Company/Organization: _____

Mailing Address: _____

_____
               (City)   (State)   (Zip Code)

Telephone Number (include area code): _____

Brief description of your products and/or services:

Name(s) of individual or team members, with titles and job descriptions. If the nomination is for a team, designate the team leader. Also, include the telephone number of a contact person.

Does your organization have a formal quality improvement program?

_____ Yes   _____ No

If yes, describe briefly.

Does your organization have an internal quality recognition program?

_____ Yes   _____ No

If yes, describe briefly.

Does your organization train workers how to improve quality using SQC or similar techniques?

_____ Yes   _____ No

If yes, describe training techniques briefly.

Figure 7.1. *(continued).*

Describe – in 750 words or less on double spaced, typed, separate pages – the accomplishments of the individual or team. The statement should include what was improved, how it was improved, how the improvement was measured, and the influence that person or team has had on product/service improvement. Is the nominee being used as a role model for the organization? The story of the individual's or team's accomplishments must be complete in the 750-word statement. Support materials detailing techniques and results may be submitted but they must not exceed three pages.

_____

Signature

_____

Title

Figure 7.1. *(continued).*

Nomination Number

Scoring, questions 1 through 8: 1= low, 10 = high.   **Score**

1. REPRODUCIBILITY OF THE PROCESS
   Was the improvement the result of a systematic
   quality-oriented process?

2. BREADTH OF EFFECT
   Does the improvement have significant effect beyond
   the immediate performance of the nominee?

3. GENERALITY OF APPLICATION
   Is the improvement specific, with application primarily
   to the work of the nominee, or does it have application
   to other operations of the organization?

4. MAGNITUDE
   How large an effect has the improvement had on the
   organization?

5. QUANTIFICATION
   Is the improvement measured or measurable by a
   quantifiable standard?

6. SUITABILITY
   Is the standard appropriate for judging the value of the
   improvement?

7. EVIDENCE
   Does the standard employ hard factual data to measure
   the improvement?

8. BENCHMARK
   Is there a benchmark that permits comparison with other
   improvements?

Scoring, question 9: 1= low, 20 = high.

9. ROLE MODEL
   Does the nominee's accomplishment demonstrate a
   commitment to quality that provides an example others
   in the organization can follow, regardless of the nature
   of their jobs?

Total Score

Figure 7.2. The Quality Cup score sheet.

Chapter 8:

The European Quality Award

JOHN A. GOLDSMITH

On September 15, 1988, the presidents of 14 major European companies signed a letter of intent to establish the European Foundation for Quality Management (EFQM). In this letter the following statement was made.

> The foundation will develop specific awareness, management education, and motivational activities in close cooperation with other European organizations and relevant institutions of the European Economic Community. The foundation will recognize achievements in quality management by establishing a European Quality Award. It will generate publicity to promote its objectives.

The idea for the award was born in September 1988 and four years of communication and hard work passed before it was announced publicly by Martin Bangemann in October 1991 and the first award was made in October 1992.

The brochure which launched the award scheme in January 1992 named the award sponsors as the European Commission, EFQM, and the European Organization for Quality. A steering group, managed by an officer of EFQM and chaired by a member of the EFQM Executive Committee, which included representatives of the other two sponsors, planned and managed the development of the award.

During the early stages of the development of the award there was a clear distinction between Eastern and Western Europe and the award was targeted at Western European organizations. The idea was, and remains, that the European Quality Award will be awarded to a number of companies that demonstrate excellence in the management of quality as their fundamental process for continuous improvement. To receive the award, applicants must demonstrate that their approach to TQM has contributed significantly to satisfying the expectations of customers, employees, and others with an interest in the company for the past few years. The European Quality Award is awarded to the most successful exponent to TQM which submitted an application for the year in question.

Development

The experience behind the major European national quality award and those from America and Japan was examined and used substantially during the formation of the detailed rules and procedures. The knowledge and experiences of those familiar with the MBNQA were particularly useful, although there are significant differences between the forms of the Baldridge Award and European Quality Award.

At an early stage it was decided that the award would embrace all the activities of any applicant, including business results and impact on society. However, it was recognized that the award should not be prescriptive and should not lay down detailed rules for corporate organization. The major criteria to be assessed were decided on following detailed discussion involving EFQM, the other sponsors, and outside bodies such as academic institutions. Nine criteria were selected which were split into enablers and results.

It was realized that assessment would present a major challenge and, after deep thought, a mechanism based on self-appraisal followed by validation of that appraisal, was chosen. A steering group of EFQM but which included representatives of the other sponsors, managed the development of the award, and the development process took approximately two years. The major effort was, of course, in constructing the criteria and assessment process but significant efforts were also used to arrive at eligibility rules and the nature of the trophy itself.

The design eventually chosen for the trophy is made of identical interlocking pieces of black corian, locked together by six elements made of silver plate. The sphere is an idealized representation of the total quality approach. Each activity in an organization is represented in the way all the components interact and fit together in such a way as to construct the universally accepted perfect shape, the sphere.

It was also decided that the award process itself would be subject to continuous review to ensure that all opportunities for improvement were grasped.

This may mean that some of the information in this chapter will be superseded as the award develops. Information presented is based on the model used for the 1992 award.

The criteria selected and the percentages of the total mark allocated to each are as follows (see Figure 8.1).

1. **Enablers (50%)**
   Leadership (10%)
   People management (9%)
   Policy and Strategy (8%)
   Resources (9%)
   Processes (14%)

2. Results (50%)

People satisfaction (9%)
Customer satisfaction (20%)
Impact on society (6%)
Business Results (15%)

Enablers and results are valued equally with the highest individual criterion being customer satisfaction, followed by people (management + satisfaction), with business results being the third most highly valued criterion.

Business results embrace the company's continuing success in achieving its financial targets and objectives together with its other business targets and objectives. The assessment is based on the company's self-appraisal information on each criterion, approached in a manner relevant to the company's business (see Figure 8.2).

For the enablers, information is required on how the company approaches each criterion with respect to defined areas and also the extent to which the approach has been deployed throughout the company.

For the results, information is required on the key parameters the company uses to measure results and data in the form of trends over three or more years for each parameter chosen. This will include the company's own targets and the relevance of those targets to all with an interest in the company together with actual results obtained against each target. Where appropriate, the performance of competitors and best-in-class organizations should be included.

Figures 8.3 and 8.4 show the two charts used by the assessors in scoring the applications.

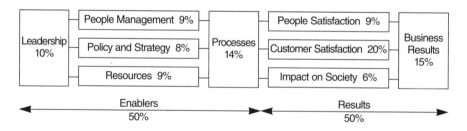

Figure 8.1. The criteria and percentages of the total score for the European Quality Award.

The application process for the award requires than you present your own company's performance across a range of specific areas relating to each criterion. The following criteria are therefore written in nonprescriptive terms, which allows you the freedom to put into your application self-appraisal information which is relevant to your business situation.

For the enablers criteria, information is required on

1. How the company approaches each criterion. Each criterion is covered by a range of specific areas and you should provide concise and factual information about each of these areas.

2. The extent to which the approach has been deployed – vertically through all levels of the organization and horizontally to all areas and activities.

For the results criteria, information is required on

1. The key parameters your company uses to measure results.

2. For each key parameter, data is required. Ideally this will be in the form of trends over three years or more. The trends should highlight

   A – your own targets
   B – the relevance of parameters to all groups with an interest in the company*
   C – your actual performance
   D – the performance of competitors (where appropriate)
   E – the performance of best-in-class organizations (where appropriate)

3. The extent to which the parameters presented reasonably cover the range of your company's activities. The scope of the results is an important consideration for the assessors.

In the case of business results, for the financials, data may be presented in the form of an index rather than absolute terms to avoid disclosing sensitive information.

To allow assessors to make comparisons more easily, it would be helpful if you would provide a single chart, for each key parameter, showing trends of A, C, D, and E. A brief commentary demonstrating understanding of significant features of the data presented is also desirable.

For each criterion you should present trend data relating to relevance of the key parameters on a single chart with a commentary on significant features exhibited by the data presented.

In the early years of the award, historic data in relation to reviews (enablers) and trends (Results) may be unavailable. Also some aspects of the criteria may not be relevant to your company. In these cases, you should not be deterred from making an application for the award. Assessors will make allowances for incomplete data, provided an explanation is offered.

* Relevance relates to the relative importance of the parameters measured in each of the four results areas. Thus in the case of customer satisfaction, information is required on the relative importance, as perceived by customers of the parameters used to measure customer satisfaction. Similar data are needed for people satisfaction. Impact on society and business results relevance, in the sense described may change over time – again, trend data are required.

Figure 8.2. Assessment criteria for the European Quality Award.

The assessor scores each part fo the enablers criteria on the basis of the combination of two factors.

1. The degree of excellence of your *approach*

2. The degree of deployment of your *approach*

APPROACH	SCORE	DEPLOYMENT
Anecdotal or nonvalue-adding.	0%	Little effective usage.
Some evidence of soundly based approaches and prevention-based systems. Subject to occasional review. Some areas of integration into normal operations.	25%	Applied to about one quarter of the potential when considering all relevant areas and activities.
Evidence of soundly based systematic approaches and prevention-based systems. Subject to regular review with respect to business effectiveness. Integration into normal operations and planning well established.	50%	Applied to about half the potential when considering all relevant areas and activities.
Clear evidence of soundly based systematic approach and prevention-based systems. Clear evidence of refinement and improved business effectiveness through review cycles. Good integration of approach into normal operations and planning.	75%	Applied to about three quarters of the potential when considering all relevant areas and activities.
Clear evidence of soundly based systematic approach and prevention-based systems. Clear evidence of refinement and improved business effectiveness through review cycles. Approach has become totally integrated into normal working patterns. Could be used as a role mode for other organizations.	100%	Applied to full potential in all relevant areas and activities.

For both Approach and Deployment, the assessor may choose one of the five levels (0%, 25%, 50%, 75%, or 100%) as presented in this figure, or interpolate between these values.

Figure 8.3.   Scoring guidelines used by the assessors for the enablers category.

The assessor scores each part of the results criteria on the basis of the combination of two factors.

    1. The degree of excellence of your *results*.

    2. The *scope* of your results.

RESULTS	SCORE	SCOPE
Anecdotal.	0%	Results address few relevant areas and activities.
Some results show positive trends. Some favorable comparisons with own targets.	25%	Results address some relevant areas and activities.
Many results show positive trends during at least three years. Favorable comparisons with own targets in many areas. Some comparisons with external organizations. Some results are caused by approach.	50%	Results address many relevant areas and activities.
Most results show strongly positive trends during at least three years. Favorable comparisons with own targets in most areas. Favorable comparisons with external organizations in many areas. Many results are caused by approach.	75%	Results address most relevant areas and activities.
Strongly positive trends in all areas during at least five years. Excellent comparisons with own targets and external organizations in most areas. Best in class in many areas of activity. Results are clearly caused by approach. Positive indication that leading position will be maintained.	100%	Results address all relevant areas and facets of the organization.

For both Results and Scope, the assessor may choose one of five levels (0%, 25%, 50%, 75%, or 100%) as presented in this figure, or interpolate between these values.

Figure 8.4. Scoring guidelines used by the assessors for the results category.

Eligibility

Winners of the award will be leading European companies, subsidiaries, or divisions which provide models of quality for others to follow. All applicants must be able to demonstrate at least five years of significant commitment to Western Europe with at least 50 percent of the applicant's business operations having been conducted in Western Europe during that period.

If a company employs less than 500 people, then the company as a whole must apply, otherwise a subsidiary or a division can apply in its own right providing that it is a clearly differentiated business. This means that it cannot be performing only support activities within a company. Its customers must perceive it as a business in its own right and not, for example, as part of a chain. At least 50 percent of its customer base must be free of direct financial and line control by its parent company. No more than three entities within a parent company can apply for the award in the same year.

Most businesses may apply and these include

- Publicly or privately owned organizations
- European and foreign owned organizations
- Joint ventures
- Incorporated firms
- Partnerships
- Sole proprietorships
- Holding companies

Those which may not apply include

- Local, regional, national and governmental agencies
- Not-for-profit organizations
- Trade associations
- Professional societies

The eligibility criteria are clearly defined at the beginning of each application period but, over a period of time, will evolve as experience grows and political changes occur.

The Award Process

The timing of the process is determined by the date of the award ceremony – usually in October – and submission of the application and the assessment process. Applications are submitted during the first quarter in the defined format as explained in the application form. Recommendations are that the application should

- Be no more than 75 A4 pages – loosely bound – in English (although original information in the applicant's own language may be included provided it is cross-referenced).

- Include self-appraisal material (the majority of the application) and fees.

On receipt each application is assessed by up to six assessors who will have undergone the same training course to maximize consistency. The application is scored on the basis of the self-assessment information provided. On the basis of the assessment, the jury decides which applicants are visited for further assessment and verification of the self-assessment information. The jury is a high-level body and each member is eminent in the field of quality and management in Europe.

After the jury has made its decisions the European Quality Awards are presented at the annual EFQM forum during a formal, high-level ceremony. Award winners are allowed to use the award symbol in promotional material and are expected to participate in meetings arranged to enable a broad spectrum of European businesses to be exposed to the procedures and arrangements necessary to achieve the standards of the winner.

The overall timetable for 1992 was

- Applications received by April 3
- Marking by teams of assessors during April and May
- Site visits arranged during June and taking place in July and August
- Final decisions taken in September
- Presentations given in October

The training for the assessors and jurors is intensive and has to be completed for the assessors by the end of March and for the jurors by the end of May. This means that the selection processes have to be completed well before these dates and the search for assessors starts well before the end of the previous year.

Unlike many other awards, the European Quality Award is truly international and assessment has to include provision for multilingual documentation review and site visits.

It is a new award which has been introduced with both political and major industrial support and it will evolve and strengthen during the coming years provided that the genuinely open and international approach is maintained.

One of its main objectives is to increase the visibility of total quality concepts and practices in Europe, thereby encouraging a wide range of businesses to adopt these concepts and practices with the overall result of improving European competitiveness in the world market. This can only happen on the ground within the individual nations of Europe and, it is hoped, and to some extent already being seen, that national quality awards will be structured using the European Quality Award as a model. Structured but open discussion with other organizations which operate strongly at the national level will be necessary to bring this about and dialogue has already begun. Due to the efforts of many people, mainly inside but also outside EFQM, the European Quality Award has been launched successfully. However, it will need broad support across Europe if it is to continue to play a major role in the European quality improvement process.

Chapter 9:

The British Quality Award Scheme

Roy Knowles

The British Quality Award Scheme was launched in 1984 by the British Quality Association (BQA). The scheme was introduced to encourage total quality improvement in commercial, industrial, and other corporate organizations based in the United Kingdom (U.K.). Since the start of the scheme it has been organized, administered, and totally funded by BQA and is completely independent of government or private authority influence and control.

Awards under the scheme are presented annually to groups (which may range from individuals to major organizations in their entirety) that are judged to have made the most significant improvement to the standard of quality of a product, process, or service. In the eight years ending in 1991 there have been 18 award winners representing enterprises as diverse as vehicle manufacture, chemical production, nuclear power plant, general engineering, electronic component and equipment, steelworks coke oven operation, food processing, and office services. In addition, more than twice this number of organizations have been highly commended. Plans are well advanced to continue the existence scheme at least through 1992 and 1993, but discussions are also taking place at the instigation of Her Majesty's (HM) government on the possibilities of developing a higher profile award scheme for the long-term future.

Any group of individuals or corporate body based in the United Kingdom may be nominated for an award, on submission of a written entry. These entries are judged by an independent panel of judges voluntarily offering their services and comprising experts from industry, commerce, and academe, appointed annually by the BQA Board of Management. The initial judging process covers an assessment of the written entries, and this is followed by a visit or visits by representative judges to the site of shortlisted entries. The adjudication criteria are primarily based on

1. A significant improvement in quality achieved over the previous four years in product design and/or manufacture of a product, or planning and/or operation of a service, or development and/or operation of a process.

2. A proven process of continuous improvement, measured by some or all of the following.

   a. Satisfying customer's requirements at a competitive cost

   b. Motivation and education of personnel to improve quality standards

   c. Better product or service performance

    d. Technological innovation to improve quality standards

    e. Quantifiable commercial or operational success

    f. A substantially higher standard than that prevalent in the relevant business sector, following the improvement in quality

3. An emphasis on actual achievement, and the contributions made by the group in the total context of the organization in which it takes place, and backed by evidence of a well-conceived and organized plan that resulted in quality improvement.

4. The overall dedication to quality pertaining within the parent organization: its management structure and the chain of responsibility and communication links between the nominated group and other departments, outside suppliers, and customers.

The annual program for the scheme requires entries to be submitted by May 1, in the relevant year, the judging process takes approximately four months, and the awards are presented at the annual dinner of BQA in November associated with the major national Quality Day event. Each award winner receives a trophy, in the form of a bronze urn to a design specially commissioned by BQA from the Royal College of Art, together with an illuminated certificate incorporating a citation on the subject of the particular quality improvement for which the award has been granted. The winners of awards are authorized to use the British Quality Award logo on printed and promotional material directly related to the specific subject of the award.

To a very large extent the operation of the scheme depends on the voluntary contribution of time and effort made by the panel of judges and supporting expert assistance given by individual members of BQA. There are, however, administrative costs involved, and these amount annually to a sum in excess of £20,000 met from BQA funds, which in turn rely mainly on members' subscriptions. No entry fees are charged, neither are any charges made for handling documentation or for the visits made by the judges.

BQA's experience in operating the award scheme has highlighted a number of factors which influence the benefits and problems associated with all such schemes. The link between granting awards and any immediate material or financial gains on the part of recipients is, to say the least, tenuous and difficult to identify. Reaping commercial rewards from quality improvements is necessarily a long-term process, requiring continuous dedication and a clear understanding that it may take years to attain the objectives of competitiveness and higher market share.

It is undeniable that the most significant improvements in quality performance achieved in recent years have been mainly motivated by a belated recognition on the part of companies that business survival is dependent on competing effectively to meet the increasingly high quality standards demanded in world markets. However, corporate bodies, like individuals, seek praise for their achievements and encouragement to promote further improvements in addition to commercial success. Award schemes, independent audits, and other forms of acknowledging progress have an important role to play in motivating organizations to continue the quality improvement process. They do not directly create better quality but they do give recognition to progress already made and provide a baseline from which further advances can be made.

There is, of course, a real danger that the formalized recognition of successes by awards or certification may be interpreted as reaching a final goal, an end in itself. It is of vital importance that top managers and, from them, the whole management structure, understand that each award represents only a single step in meeting what is a never-ending challenge. The attainment of quality at a level to meet competition presents a moving target requiring an ongoing commitment to reach ever-higher standards. It follows that all schemes for awards and certification should be dynamic with a continuing program for raising criteria levels if they are to successfully contribute to maximizing quality performance. Sadly there is a trend on the part of some award sponsors and certification authorities to develop bureaucratic procedures far too rigid to accommodate changes as they become necessary or desirable. It is essential that procedures should be designed in such a way that they remain sufficiently flexible to evolve with advancing technology and the development of more intense international competition.

The demands facing different organizations, countries, and multinational trading groups vary considerably, and it would be counterproductive to attempt to standardize management systems or the means of measuring and judging their performance. Historical, cultural, and social differences, in addition to purely commercial considerations, require dissimilar approaches to quality improvement. It is, however, possible to establish some common ground in assessing the relative merits of all quality improvement activities – for example, the results in achieving improved standards of products and services, better levels of productivity, extending technological benefits to a larger proportion of the world population, higher returns on investment and higher levels of job satisfaction, more socially and environmentally acceptable progress – in short the creation of greater real wealth. In this respect quality award schemes and other similarly intended projects can be helpful in charting a course in the right direction. Certainly some criticism of their existing form and operation can be justified, but if made constructively can form the basis for progressive improvement. None of these schemes can provide a panacea for dealing with quality problems, but they can and do make a valuable contribution and deserve our ongoing support.

Update

In February 1992, the Department of Trade and Industry, acting on behalf of HM government, set up a committee under the chairmanship of Denys Henderson (chairman of Imperial Chemical Industries) to consider the feasibility of a new U.K. total quality award scheme. This committee completed its work in August 1992 with the publication of a consultative document which is now in circulation for comments. It includes proposals for a major national initiative to introduce an award scheme similar to the U.S. Baldrige Award and conforming with the Europeanwide scheme initiated by the EFQM in 1992. If the proposals are accepted by HM government, then the new scheme will be operated on a scale significantly larger than the existing British Quality Award, which it will replace from 1994 onwards.

The 1992 annual dinner has been arranged to take place in London on November 11, 1992, at which the guest of honor, Michael Heseltine, MP president of the Board of Trade, will be making the presentations to the 1992 award winners. It is anticipated that he will also announce HM government's intentions regarding the future of quality awards in the United Kingdom. Details of any new award scheme, and decisions on how and by whom it will be operated, are however not expected until later in the year.

Chapter 10:

The Scandinavian Quality Awards

KERSTIN JÖNSON

ASBJØRN AUNE

Two Types of Quality Awards

In the Scandinavian countries Denmark, Norway, and Sweden there are two types of quality awards.

1. Quality awards with criteria for excellent quality management to be used for self-evaluation and action programs with the possibility to win a prize

2. Quality awards to reward remarkable quality efforts and results

The general purpose of both types is to promote and support the interest for quality and TQM in order to strengthen national competitiveness. In all Scandinavian countries the interest in TQM is increasingly influenced by the global competition and the related needs for efficient organizations with ability to meet the total needs of all stakeholders. The value of having models for actions has become obvious throughout the world. The awards built on criteria have an important role in building models. At the same time, it is of value to have good examples. Here both the awards built on criteria and a connected possibility to win the prize and the rewarding awards have a role. In the Scandinavian countries the awards have existed for different periods. All of them undergo development in line with worldwide trends in quality.

Finland has close connections with Scandinavian countries. In this country there is also a development of a new quality award based on a model for excellent quality management with influence from other awards.

The Danish Quality Award

In Denmark the Danish Society for Quality is responsible for the existing national quality award, which has been given since 1985. The rules and criteria have been the same up until now. New criteria, however, are being considered by the Board of the Danish Society for Quality.

In addition to this, there is an investigation going on in Denmark to find out if there are good reasons to establish another Danish Quality Award to replace a number of existing awards. The final decision doesn't exist yet. The image will be an important factor to take into consideration in this decision. If there will be a common Danish Quality Award or if the Quality Award of the Danish Society for Quality will remain, the criteria will be developed or changed to be in line with the actual criteria of the MBNQA and the European Quality Award.

The purpose of the existing Quality Award of the Danish Society for Quality is

- To recognize remarkable efforts in quality management in order to stimulate other organizations to work with quality management in a systematic way by establishing and implementing quality systems, quality improvement programs, development of new methods, motivation of management and employees, and so on

- To increase the interest in the purpose and activities of the Danish Society for Quality

The rules say that

- The award is given for remarkable efforts in quality management in its broadest sense.
  - Systematic work
  - Involvement of employees
  - Product improvements
  - Cost reductions
  - Openness concerning quality management methods and results

- The award is given to Danish companies, organizations, and institutions. The award cannot be given to a single person.

- The award can be presented once a year and normally only to one company, organization, or institution.

- The winner is obliged – in connection with the award ceremony or in another occasion within six months from the date of the award – to give orientation about the efforts that led to the winning of the award. The orientation is given in a form decided by the Board of the Danish Society for Quality.

- The award is given in the form of a diploma with documented motives.

The award process from the nomination to the presentation of the award is surrounded by some rules.

- The Board of the Danish Society for Quality nominates in February of each year a Nomination Committee of four members, one of which must be a member of the board.

- Every member of the Danish Society for Quality may present to the Nomination Committee a proposal concerning a winner. The proposal shall be backed by reasons. The proposals shall be presented at a latest date decided by the board.

- The Board of the Danish Society for Quality makes the final decision about the winner based on recommendations from the Nomination Committee. The decision by this board is without appellation.

- The award ceremony takes place in connection with the annual general assembly of the society.

- The Board of the Danish Society for Quality has the ability, from time to time, to deviate from the rules when required by the circumstances.

Five Winners So Far

Since 1985 five winners have been given the award. The companies that have won are

- 1985 – Danfoss A/S
- 1987 – A/S Ernst Voss Fabrik
- 1989 – Viking A/S
- 1990 – Grundfos A/S
- 1991 – Danochemo A/S

The Norwegian Quality Award

The history of the Norwegian Quality Award goes back to the early 1970s when the Norwegian Society for Quality (then called the Norwegian Society for Quality Control) accepted the idea of Asbjørn Aune to create the Norwegian Society for Quality Control's Quality Award. The first award – for the year 1974 – was given to SIEMENS A/S plant in Trondheim at the society's yearly Quality Days – a two and a half day conference – in May 1975.

The purpose of the award was

> To reward outstanding efforts in the fields of quality management/quality control (QM/QC) in order to stimulate companies, institutions, corporations, and individuals to use existing and

develop new methods to create more efficient QM/QC and to increase the interest in and publicity around the society's purpose and work.

The reward could be given to Norwegian companies, nonprofit institutions and individuals, one winner each year. Suggestions could be made by the members of the society and the final decision was made by the board. The recipients of the award in the period 1974 – 1990 are listed in Figure 10.1.

The New Award

Looking back it seems clear that the Norwegian Society for Quality (Control) introduced the prize too early. The first years the publicity around the prize was modest, but the interest grew through the 1980s and, in 1989, the time had come to change the name to the Norwegian Quality Award. In the meantime the society had dropped the word "Control" from its name. In the beginning, the president of the society and later a cabinet minister handed out the first 17 awards. In 1992 the president of the Norwegian parliament agreed to perform the ceremony. He/she is next to the Norwegian king in ceremonial rank.

With the change in name came a change in statutes. Today the purpose of the award is

- To reward outstanding efforts in the field of quality management and use of QC principles, in order to stimulate companies, (nonprofit) institutions, and organizations to a continuous effort to improve the quality of their products and services to the benefit of their customers

- To create an opportunity for companies, institutions, and organizations to have their standings in the field of quality to be evaluated by experts – as a part of their future work on quality

- To increase the interest in and publicity around the society's purpose and work

In the future the companies, institutions, and organizations have to apply themselves for the award. It will be possible to win the award more than once. The award is handed out once a year, and there might be more than one winner. There might also be no winner. During the year following an award, the winner(s) have to share their experiences in articles and papers at conferences and be positive to requests from others to visit them.

1974:   SIEMENS A/S, Trondheim
          Large electrotechnical

1975:   Siv.ing. Ovlav Selvåg, Oslo
          Personal, large, housing construction

1976:   Bjerke Maskinfabrikk A/S, Kongsvinger
          Small, mechanical

1977:   Frank Mohn A/S, Bergen
          Medium, mechanical

1978:   Forsvarets Felles Materielltjeneste (FFMT), Oslo
          The Office of Quality Assurance – Norwegian Armed Forces

1979:   A/S Helly Hansen, Moss
          Medium, textile

1980:   O. Mustad & Sønn, Gjøvik
          Medium, fish hooks

1981:   A/S Nora-Sunrose Konservesfabrikk, Hamar
          Medium, food

1982:   Notodden Elektronikk A/S, Notodden
          Small, electronics

1983:   Aker Stord A/S, Stord
          Large, offshore structures

1984:   Statoil Gullfaks A-Prosjekt
          Large, oil-field development project in the North Sea

1985:   EB-Nera, Bergen
          Medium, electronics

1986:   MITECH AS, Røros
          Small, electronics

1987:   Toro Næringsmiddelindustri, Rieber & Søn A/S, Bergen
          Large, dried food

1988:   Sivesind Møbelfabrikk A/S, Gjøvik
          Small, kitchen furniture

1989:   Elopak A/S, Spikkestad
          Medium, dairy products packing

1990:   The Rasmussen Group
          Large, shipowners

The award for one year is handed out the following year. *Small, medium* and *large* are defined as follows: Small – less than 100 employees, medium – 100-500 employees, and large – more than 500 employees.

Figure 10.1.   Recipients of the Norwegian Society for Quality's Quality Award 1974 – 1990.

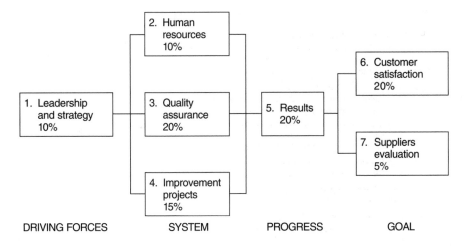

Figure 10.2. The criteria for the Norwegian Quality Award.

The Criteria, the Scores, and the Selection Process

A winner must be able to show that its work in the quality field has resulted in increased customer satisfaction and improved economical and environmental results. See Figure 10.2 for the award criteria and Figure 10.3 for a breakdown of the scoring guidelines.

The selection process includes these rules.

1. Applications must be received by the secretariat of the Norwegian Society for Quality.

2. The applications must be evaluated by two or three examiners.

3. Decision point: Companies not satisfying minimum requirements will be informed and given comments of strengths and weaknesses.

4. The Award Committee will evaluate the remaining applications and inform companies which will not be visited. The committee will also comment on strengths and weaknesses.

1.0	**Leadership and strategy**	**10 points**
	1.1 Top management commitment and involvement	4 points
	1.2 Goals, strategies, and implementation	4 points
	1.3 Communication and information	2 points
2.0	**Use of human resources**	**10 points**
	2.1 Goals for improvement of the human resources	2 points
	2.2 Employees participation, responsibilities, and authority	2 points
	2.3 Education and training of employees	2 points
	2.4 Recognition of employees	2 points
	2.5 Working environment and attitudes	2 points
3.0	**Quality assurance of goods and services**	**20 points**
	3.1 Process control and surveillance	10 points
	3.2 Documentation	6 points
	3.3 Quality of suppliers	4 points
4.0	**Improvement projects**	**15 points**
	4.1 Product and process development	5 points
	4.2 Continuous improvement	5 points
	4.3 Quality evaluation	5 points
5.0	**Results**	**20 points**
	5.1 Economical planning and result improvement	8 points
	5.2 Market position and development	8 points
	5.3 Social responsibility	4 points
6.0	**Customer satisfaction**	**20 points**
	6.1 Understanding and identifying customers' expectations and requirements	4 points
	6.2 Organization and control of customer relations	4 points
	6.3 Measurement of customer satisfaction	7 points
	6.4 Handling of complaints	5 points
7.0	**Supplier's evaluation**	**5 points**
	7.1 How the suppliers view the company	5 points
		Total 100 points

Figure 10.3. Scoring guidelines for the Norwegian Quality Award.

5. The examiners will make site visits to the finalists.

6. The final decision will be made by the Board of the Norwegian Society for Quality.

The board of the society will elect examiners who, as a minimum, must be lead auditors approved by the society with additional training in the award criteria.

Experiences and Expectations

There have been no systematic studies of the effects – internal or external – to the award-winning companies. However, what can be stated is that a majority of the companies today are well-known in Norway both for the quality of their products and their quality management programs. Some of them were, and still are, among the international leaders in their fields.

It should be mentioned that the Norwegian Society for Quality in 1984 inaugurated a Quality Circle Award – which each year since has been given to one Norwegian quality circle.

The Swedish Quality Award

The Swedish Quality Award started in 1992 – a year earlier than planned. This depended on the fact that the Swedish industrial leaders were concerned about the competitiveness of Swedish industry and saw the value in having an award giving everybody inspiration and a model for quality and quality management efforts. The hope among all supporters is that this award will create a special quality culture on a national level. The intention is that both production and service companies, including banks, insurance companies, as well as governmental bodies will be attracted by the award and the criteria reflecting a total approach to quality and will take part in the competition.

The History Behind the Swedish Quality Award

The Swedish Quality Award has its background in a common national effort for increased quality consciousness. The inspiration for this effort grew from three different directions.

1. The interest to develop the employees in all industrial organizations in their ability to take broader responsibilities in their work and establish job ownership

2. The renewal of the public sector

3. The recognition of the importance of quality for the industrial development

In November 1986 a Swedish offensive for quality started with the motto The Right Way. This offensive was driven by a National Committee for Swedish Quality backed by the government, the industry, the unions, and other organizations. The Royal Highness, Prince Bertil, was the chairman of the board. The purpose of the offensive was to renew the company and service culture in private and public activities. The goal was to get continuous changes and improvements of the quality of goods, services, and work.

After having run this offensive for some years a need for a converging, permanent, and driving institution matured. In 1990 the Swedish Institute for Quality was established in order to meet these needs. This institute is sponsored by the government and private and public companies and organizations. One of the tasks given to this institute has been to establish a Swedish national quality award in class with the Deming Prize in Japan and the MBNQA in the United States. This quality award should be a model of value for all sectors of the society.

International Influences

Sweden has a broad international orientation and is very much dependent on its export industry. Therefore it is of vital importance to work according to state-of-the-art quality and to make efficient use of already existing models. Based on this, the Swedish Quality Award, developed in 1992, is very much a copy of the MBNQA with an adaptation to Swedish conditions, traditions, and ambitions. Before the award was launched, a preparatory test was performed to test the applicability of the MBNQA criteria in Sweden. The outcome of this test was, among other things, that all involved organizations liked the broad approach to quality. At the same time, they realized the challenge in getting a high score.

The Fundamental Values

The 13 fundamental values defined for the Swedish Quality Award are

- Customer orientation
- Leadership
- Participation by all
- Competence
- Long-range outlook
- Public responsibility
- Process orientation

- Preventive activities

- Continuous improvements

- Learn from others

- Fast reactions

- Decisions based on facts

- Cooperation

The Model and the Criteria

The Swedish model is based on seven groups of criteria like the MBNQA (see Figure 10.4). The relationship between the criteria is shown in Figure 10.5. Scoring guidelines are summarized in Figure 10.6.

Two Categories

The Swedish Quality Award 1992 is open for two categories:

- Goods or service producers with 200 or more employees

- Goods or service producers with less than 200 employees

The producer should form an independent unit covering all award criteria possible to verify in Sweden.

The winner of the Swedish Quality Award in 1992 is not allowed to participate in 1993 and 1994. The final decision who is allowed to participate is taken by the Swedish Institute for Quality. The participation fees for the different categories are SEK 2,5000 and SEK 7000 respectively.

All parties involved in the application and evaluation process have defined obligations and commitments. These are clearly defined for the Swedish Institute for Quality, the participants, the winners, the examiners, and the judges. One of the most important is the winners' commitment to disseminate knowledge about their own quality improvement work to all interested.

The evaluation process is established so that each participant is ensured an unbiased, fair, and competent evaluation. The two competition classes are judged separately. The evaluation process is done in several steps and ends with a feedback report (see Figure 10.7).

1	**Leadership**	**90 points**
	1.1 Senior executive leadership	45 points
	1.2 Leadership for continuous quality development	25 points
	1.3 Public responsibility	20 points
2	**Information and analysis**	**80 points**
	2.1 Scope and management of facts about quality	15 points
	2.2 Comparisons with competitors and quality leading organizations	25 points
	2.3 Uses of facts	40 points
3	**Strategic quality planning**	**60 points**
	3.1 The way of working with strategic planning	35 points
	3.2 Plans and objectives for quality and efficiency	25 points
4	**Development, involvement, and participation of employees**	**150 points**
	4.1 Development of employees	20 points
	4.2 Involvement and participation of employees	40 points
	4.3 Education and training	40 points
	4.4 Recognition and stimulation of quality work	25 points
	4.5 Work satisfaction and work environment	25 points
5	**Quality in operational processes**	**140 points**
	5.1 Development and design processes	35 points
	5.2 Production and distribution processes	30 points
	5.3 Support processes	25 points
	5.4 Supplier cooperation	20 points
	5.5 Management of environmental issues	15 points
	5.6 Assessment processes	15 points
6	**Operational results**	**180 points**
	6.1 Results – Goods and services	65 points
	6.2 Results – Design, development, production, and distribution processes	40 points
	6.3 Results – Support processes	25 points
	6.4 Results – Supplier cooperation	30 points
	6.5 Results – Environmental impact	20 points
7	**Customer satisfaction**	**300 points**
	7.1 Management of customer relations	65 points
	7.2 Commitments of customers	15 points
	7.3 Customer satisfaction determination	35 points
	7.4 Customer satisfaction results	75 points
	7.5 Comparisons with competitors	75 points
	7.6 Future requirements and expectations of customers	35 points

Total points   **1000**

Scoring in three dimensions

The scoring system also has its base in the MBNQA scoring system. The scoring is based on the evaluation dimensions

1. Approach
2. Deployment
3. Results

Figure 10.4. The Swedish Quality Award criteria.

The Swedish Quality Award will be presented by the king of Sweden to the winners of the award. The award ceremony is planned to take place in the ancient House of Commerce in Gothenburg, the second largest city of Sweden, in which the Swedish Institute for Quality is located.

The first year for the Swedish Quality Award, 13 companies participated in the competition to win the prize. Eight of these are organizations with less than 200 employees. There were difficulties to predict how many would participate and to know the need for examiners. A qualified estimate resulted in that about 60 examiners were nominated and have got special education and training concerning the criteria and the evaluation process.

The overall impression is that the Swedish Quality Award 1992 has become a success. Besides the 13 companies that participated in the competition, more than 1000 companies and other organizations have ordered more than 14,000 copies of the criteria and rules. These companies and organizations are now working internally to improve their quality and competitiveness.

One of the experiences gained during the evaluation process is the importance of the logic in the definition of the criteria to make them easy to understand and to answer in the application report. This is one of the reasons why there will be some changes in the criteria for coming years.

Another experience is the importance of having an efficient examination process backed by useful approaches and tools for reports. This will make it possible for highly qualified professionals to be involved in this task besides their ordinary work. This is of great importance for a small country like Sweden where, for the time being, there are a relatively small number of quality professionals with the knowledge and experience required.

Due to the fact that the first evaluation process has not yet come to an end there is no systematic knowledge and evidence gathered about the impact on the quality activities and the quality culture in Sweden. However, the Swedish Institute for Quality will, in the coming year, sponsor projects with the focus on the Swedish Quality Award. Questions to address will include How are the criteria working in different organizations, such as goods and service producers, private and public organizations, and so on. How is the improvement work driven based on the award criteria and which results have been achieved?

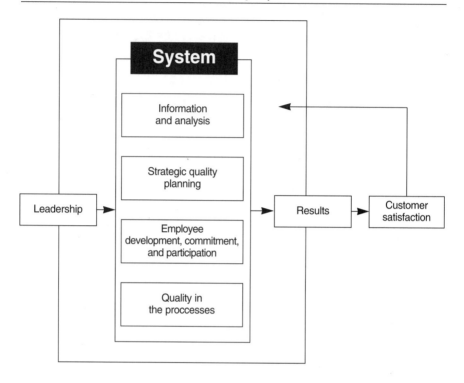

Figure 10.5. The relationship between the criteria of the Swedish Quality Award.

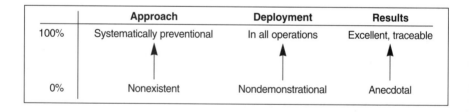

Figure 10.6. Summary of scoring guidelines for the Swedish Quality Award.

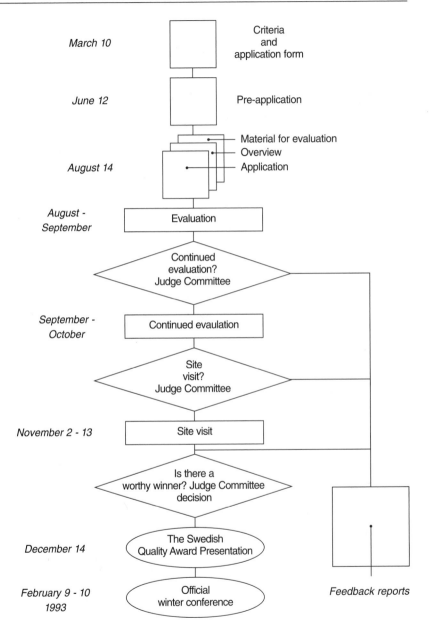

Figure 10.7.   The Swedish Quality Award evaluation process.

Chapter 11:

Prix Qualité France

ALAIN-MICHEL CHAUVEL

Set up on the initiative of the French Ministry for Industry and Foreign Trade and the French Quality Movement (MFQ), the Prix Qualité France was awarded for the first time in 1993 to the company Spirotechnique.

Is this just another item in the race for trophies and awards, so prized by companies yearning for recognition? No. The Prix Qualité France is the culmination of an experimental scheme that goes back to 1980.

At that time, the French Association for Industrial Quality Control (AFCIQ) and the magazine *L' Usine Nouvelle* jointly awarded the first industry and quality prize to two companies.

- Maubeuge Construction Automobile (MCA)
- Luis, a firm employing less than 500 people

This prize was intended to reward efforts made by companies to achieve maximum quality in their operations. The tradition has continued every year since those first two prizes, and winners have included names enjoying international prestige, and national or regional companies that deserve recognition, like Aerospatiale, Fasson, Agenieux, Rank Xerox, April, and Fabris Frères & Fils.

So why change the formula? Since 1980, quality has made a breakthrough in French business, and it seemed reasonable for the prize scheme also to evolve. The originality of the Prix Qualité France is that it is no longer a single prize awarded annually to one company, but forms one of the links in the common chain of quality so vital to the France of tomorrow. Another change is that it is now aimed exclusively at small and medium-size companies, which do not always have the means of making themselves known and being recognized at a national, much less an international, level. The award therefore helps discover and commend a model firm, mobilized around that paramount criterion of competitiveness: quality. It helps to further recognize the contribution of quality in all aspects of business management and applaud the participation of the men and women who bring about progress for their company.

Principle of Operation

Small and medium-size companies first compete at a regional level. Already in 1992, more than 1000 French companies had taken part in these regional stages. Regional prizewinners are encouraged to compete for the Prix Qualité France. Thirty-six companies were involved for the 1993 competition.

Spirotechnique, the winner of Prix Qualité France 1993, was selected from the 36 regional winners. The jury also granted three honorable mentions to outstanding runner-ups: Socat, Amsusa, and Atelier du Val d'Or.

Regional Prizes

Candidates submit details of their quality approach in their application. The jury, set up on the initiative of the Regional Division for Industry, Research & the Environment, and the MFQ regional bureau, analyzes the applications and may appoint surveyors to visit candidates.

The prize is awarded on the basis mainly of

- Quality and exemplary nature of the company approach to quality

- Performances expected and obtained, particularly as regards customer satisfaction

- The extent and exemplary nature of results obtained or expected, with particular attention to the involvement of the workforce concerned by the establishment and implementation of the quality approach

Each region awards a regional prize and honorable mentions. The winners of these regional prizes go forward for the national prize.

National Prize

Candidates selected in this way for the Prix Qualité France are examined by a national jury, set up on the initiative of the Ministry for Industry and Foreign Trade and the MFQ. A single prize is awarded, together with honorable mentions, to those companies selected by the jury.

Candidates must submit an application file. The jury, assisted by an assessment committee, analyzes and evaluates applications in order to decide which candidates will receive the assessment visit needed to obtain the prize.

The application is based principally on a limited number of easily understood criteria. This enables the head of the company, alone or with his or her management team, to make a quality diagnosis of the firm, but also to discern its strengths, and detect any gaps in the quality approach. The application asks questions such as, How does your quality approach currently relate to your corporate goals? Why and when did quality become a matter of special importance in your company? How? What were the initial targets? What are the main steps that you have taken in the matter of quality up to the present?

After this summary of information, the application focuses on eight aspects of quality (see Figure 11.1).

The marking system, used to select the winner of the national prize from among regional prizewinners, is shown in Table 11.1.

The Ministry for Industry and Foreign Trade figures show that small and medium-size businesses are increasingly concerned with quality. To assist them in their approach, regional consulting aid funds, even though these are being increased annually, are not sufficient to create the dynamic needed to reach the whole industrial fabric of such companies.

The Prix Qualité France is now a stimulus for healthy competition among companies. It was necessary, like a phoenix, to allow regeneration and rebirth to occur. Like similar European, American, and Japanese prizes, the Prix Qualité France reflects the efforts and achievements of French companies to improve the quality of their organization, and of the products and services they offer their clients.

Table 11.1.   Scoring guidelines of the Prix Qualité France.

Each of the eight aspects is marked out of 100	Multiplying coefficient	Total
1. Management commitment	1.5	150
2. Quality policy	1	100
3. Customer satisfaction	2	200
4. Quality system	1	100
5. Quality indicators and measurement; management of nonconformities, preventive action, and quality progress	1.5	150
6. Improvement of quality	1.5	150
7. Workforce participation	1	100
8. Effect of quality on company earnings	0.5	50
Total		**1000**

1. Management commitment

   1.1 How does management inspire and direct company quality policy?

   1.2 How is the management commitment made visible?

   1.3 How does management acknowledge and enhance individual and team efforts and successes?

2. Quality policy

   2.1 What is your quality policy?

   2.2 Have you expressed it formally in writing (in particular for your clients and suppliers and within the company)? Since when?

   2.3 How has quality policy been circulated? How is it made known to the whole workforce? Is it reviewed and updated?

   2.4 Has the company a designated quality manager? Since when? To whom is he or she accountable?

3. Customer satisfaction

   3.1 How do you find out and/or anticipate your clients' needs and expectations regarding quality? How do you take them into account?

   3.2 How do you measure your clients' satisfaction? How do you take the findings into account?

   3.3 How do you identify your competitors? How do you position yourself in relation to them on a quality level?

4. Quality system

   4.1 What is your quality system? How is it formalized?

   4.2 How do you define the tasks and working methods deployed to respect company quality commitments?

   4.3 How are the corresponding documents created, revised, filed, and distributed?

   4.4 How do you formalize the requirements needed to gauge the quality of service by your suppliers?

   4.5 How do you achieve progress in your quality system? More specifically, do you ensure that achievements will endure?

Figure 11.1. Questions asked on the application form for the Prix Qualité France.

5. Quality indicators and measurement; management of nonconformities, preventive action, and quality progress

   5.1 What quality indicators and associated means of measurement have you set up? Do they allow you to ensure your commitments and estimate your strengths and weaknesses?

   5.2 How do you use such measurements to achieve advances in your production methods and products (for example, corrective and preventive action, action on progress)?

6. Improvement of quality

   6.1 What actions are currently under way or scheduled in order to improve quality in your company?

   6.2 What are the quantified goals of such actions? What are the estimated deadlines for their attainment?

7. Workforce participation

   7.1 How does the workforce contribute to setting company quality targets?

   7.2 How is the workforce informed of the company's evolving quality goals and successes?

   7.3 How is the workforce involved in implementing quality actions?

   7.4 What quality training is provided?

   7.5 How do you assess the commitment of the workforce to quality procedures and goals announced?

8. Effect of quality on company earnings

   8.1 What effect does quality have on your company earnings?

   8.2 How can you evaluate the quality contribution
   - economically (for example, sales, profits, costs, market shares)?
   - to company reputation?
   - at the level of the workforce?

Figure 11.1. (*continued*).

Chapter 12:

The Italian FONTI Quality Award for Small/Medium-Size Service Companies

Tito Conti

Since a national quality award has not been defined to date in Italy, a less general but nonetheless national in scope award will be described, due to its peculiar characteristics: The FONTI Quality Award for the innovative service (tertiary) businesses. FONTI is a federation of associations (Federazione delle Organizzazioni del Terziario Innovativo) whose members generally are small/medium-size companies; but the award is open to all advanced service companies, members, and nonmembers.

The peculiarity of the award is that it has been conceived with the small/medium-size service companies in mind. Today, particularly in Europe, companies of that size are (or are in the process of becoming) very much familiar with the ISO 9000 standards; in the case of service, with ISO 9004-2. Since the effort to build the company quality system and the associated culture is considerable for small companies, it seems unwise to oblige them, as soon as they have become familiar with the ISO 9004-2 model, to convert to another model, if they want to apply for the award. For that reason, the FONTI award has taken the European Quality Award as the reference model, but only for the results part (right-end side of the model). For the left-end side, the enablers, the ISO 9004-2 quality system and processes model was taken.

ISO 9004-2 is sufficiently advanced as a model for the quality system of small/medium-size companies. It fits their needs, and they find continuity going from the level of certification to the level of the award. What in the case of certification is simply a go/no-go test, in the case of the awards becomes a scoring process, with results added to the picture (with a 50 percent weight, as in the European Quality Award).

The FONTI Quality Award is presented as an interesting example of adaptation of a generic award model to the characteristics and needs of a specific business sector.

The FONTI Quality Award

Today companies are realizing more than ever that the only way to survive in business is to focus on quality. Quality must be part of all company activities and encompass all the ways in which a company meets the needs of its customers, employees and, in general, the social and economic community in which it operates. This concept is gaining more and more consensus within innovative tertiary companies.

As a result, FONTI, the innovative tertiary federation, has decided to institute the FONTI Quality Award for companies in this important sector. The award is a recognition of the commitment and achievement in the field of quality.

Objectives of the FONTI Award

The objectives to be achieved are

- To define a standard for companies that wish to excel in quality

- To stimulate a process of growth for the culture of quality and to promote global competitiveness among innovative tertiary businesses, in the knowledge that such an improvement is also essential for the development of the country system

- To demonstrate the results obtainable with quality

Benefits of the FONTI Award

An application for the FONTI Quality Award requires each candidate company to gather and organize quality data and information for drawing up the self-appraisal document. In this way a detailed picture of quality within the company may be built up as well as the extent and application of the quality approach on all levels. Furthermore, it will provide a feedback report that will highlight the areas most in need of improvement.

When companies apply for the FONTI Quality Award, an assessment process is started up that will provide the following benefits.

1. For the applicants
   - Self-analysis based on a complete and tested model
   - Possible visits by the award jury for further assessment
   - Feedback from the commission.

2. For the winners
   - Status attached to winning the award
   - Publicity of the event
   - The right to use the FONTI Quality Award symbol for corporate literature

Eligibility Requirements

To apply for the FONTI Quality Award, companies working in the innovative tertiary field must have a minimum of 10 full-time personnel (not necessarily employees). To participate, businesses must be legally registered companies and be able to show their balance sheets for the last three years. The rules and evaluation criteria used by the FONTI Quality Award have been defined by taking into account the most commonly found challenges and difficulties facing small and medium-size companies in the innovative tertiary sector. However, larger companies are also invited to participate.

The applicant must submit to the following prerequisites.

- The proven knowledge and implementation of the rules contained in the ISO 9000 series standard (UNI-EN 29000 series)

- A real commitment by the company to apply the total quality principles to its management

Presentation of the Award

The FONTI Quality Award is assigned annually to a maximum number of five winners who demonstrate a level of excellence scored on the basis of the parameters given in the document Self-Appraisal Criteria. The evaluation and scoring of the applicants, carried out by qualified assessors, will be based on the assessment model described later in this chapter. If, during the first three years, it is deemed inappropriate to award the FONTI Quality Award to any one applicant, the award may be substituted by a special mention, an acknowledgment of the validity of initiatives adopted by the company to improve quality. The awarding of the FONTI Quality Award or the special mention is irrevocable.

Timetable

Application submission and delivery of the evaluation documents must take place between January and March 1993. The assessors' commission shall notify the applicants of the results of the documental analysis. In the event of a positive result, a site visit shall be arranged. The winners shall be notified by September 1993. Presentation of the award shall take place in November 1993.

Required Application Documents

When submitting an application, the candidate should send the following documents to the awards secretariat.

1. The application form fully filled in and signed by the chairman or the managing director of the company.

2. Eight copies of the self-appraisal report drawn up according to the instructions detailed here. The report should be

- No more than 30 A4 pages in length
- Numbered on every page
- Typewritten
- Bound

3. Application fee payment receipt.

The self-appraisal report should have the following structure.

- Title page with applicant's name, address, and date of application

- Table of contents

- A two-page introduction with a brief description of the company's history, the services offered, and the events that have most influenced the life of the company

- The self-appraisal report itself based on the self-appraisal criteria

- Any documents deemed suitable to support the report (for example, a quality manual)

Assessment Process and Presentation of the Award

The self-appraisal document submitted by the applicant will be examined by a commission that will assign points based on the assessment model used. On the basis of this assessment, companies will be selected to host a site visit to check the consistency between the company's report and actual company practice. The presentation of the awards or the special mentions will be based on these results.

The award presentation process is as follows

1. Application request

2. Delivery of the application documents

3. Assessment of the applicants

4. Selection of the applicants for the site visit

5. Site visit

6. Adjudication of winners

7. Presenting the awards or honorable mentions

8. Publicizing the names of the winners

The Award Assessment Model

Applicant selection will be carried out on the basis of the self-appraisal criteria that are based on the ISO 9004-2 standard and the European Quality Award. The award assessment model is divided into two areas.

1. Enablers, where the criteria evaluate the methods by which the company best uses its own resources for the work activity

2. Results, where the criteria evaluate the qualities the company has obtained by using those methods

Briefly, the results criteria concern what the company has achieved and is achieving. The enablers criteria concern how these results are being achieved (see Figure 12.1). The enablers criteria have been borrowed from the ISO 9004-2 standard while the results criteria relate to the European Quality Award.

Enablers

Quality System Principles

Customer satisfaction is assured only when harmonious interaction exists among management responsibility, personnel and material resources, and the quality system structure (see Figure 12.2). The company should describe how it is organized, with regard to the management of each key aspect and its improvement, according to the following outline.

1.1 Management Responsibility
Management is responsible for establishing a policy for service quality and for customer satisfaction. The successful achievement of this policy depends on the management commitment in developing and implementing a quality system. Management commitment is realized by the implementation of the following quality activities.

- Definition of policies

- Definition of objectives

- Definition of responsibilities and authorities

- Periodic reexamination of the quality system

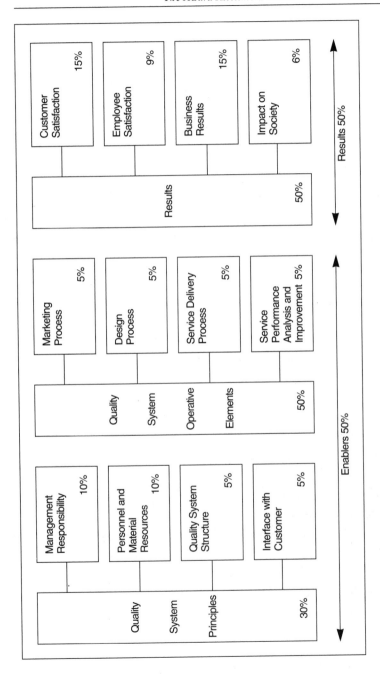

Figure 12.1. The scoring guidelines of the FONTI Quality Award.

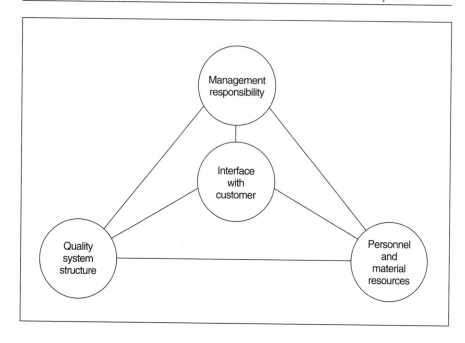

Figure 12.2.   Key aspects of a quality system.

The company should describe how these activities are carried out and how it is organized to improve them.

1.2 Personnel and Material Resources
Management should provide sufficient and appropriate resources to implement the quality system and achieve the quality objectives.

One of the determining elements in obtaining notable improvements in an innovative tertiary company's performance is represented by the quality of personnel resources. Personnel resources are managed through the following activities.

- Motivation

- Training and development

- Internal communications to ensure the correct distribution of information

- Management of full-time personnel

The company should describe how these activities are carried out and how it is organized to improve them.

The quality of a service rendered by an innovative tertiary company may be influenced by the management of the *material resources* required to carry out the service, or in support of the service itself. The company should describe how these activities are carried out and how it is organized to improve them.

1.3   Quality System Structure
The service organization should develop, establish, document, implement, and update a quality system as a means by which stated policies and objectives for service quality may be accomplished.

A quality system is achieved through the following activities.

- Monitoring of all operative process cycles that influence the quality of the service

- Activities records

- Documentation control

- Inspections of the quality system performance and efficiency

The company should describe how these activities are carried out and how it is organized to improve them.

1.4   Interface with Customer
Management should establish effective interaction between customers and the service organization's personnel. This is essential for customers' perception of service quality. Personnel in direct contact with the customer is an important information source for the actual quality improvement process. The service quality as perceived by the customer often takes place through communication with the personnel and with the service organization structure. The company should describe how it handles communications with the customer in both directions (company to customer, customer to company) and how it is organized to improve it.

Quality System Operative Elements

1.5 Marketing Process

The marketing process is composed of the following activities.

- QC in market research and analysis

- Establishing the obligations assumed by the company toward its customers

- Drawing up a service brief

- Planning service management and resources availability

- QC in advertising

The company should describe how these activities are carried out and how it is organized to improve them.

1.6 Design Process

The process of designing a service involves converting the service brief into specifications for both the service and its delivery and control. The service specification defines the service to be provided, whereas the service delivery specification defines means and methods used to supply the service. The QC list specifies the procedures for evaluating and controlling the characteristics of the service and its supply.

The design process requires

- Definition of design responsibility

- Definition of the service specification

- Definition of the service delivery specification and the related procedures (procurement QC and assurance of equipment given to the customer to use, service tractability and control of customer assets handling)

- Definition of the service control process and its specifications

- Verification that the design is consistent with the serve brief

- Validation of the service, supply, and QC specifications

The company should describe how these activities are carried out and how it is organized to improve them.

1.7 Service Delivery Process
Management should assign specific responsibilities to all personnel implementing the service delivery process.

Consequently, this entails

- The evaluation of service quality by the company

- Customer evaluation of service quality

- Recording status of the service

- Assigning responsibility for corrective action, identification and recording of nonconformity, and control of the measuring system

The company should describe how these activities are carried out and how it is organized to improve them.

1.8 Service Performance Analysis and Improvement
A continuous evaluation of the service process operations should be carried out in order to identify and actively pursue all service improvement opportunities. To implement these evaluations, management should establish and maintain a system for

- Data collection and analysis

- Statistical methods used

- Management of ongoing improvement programs of service quality

The company should describe how these activities are carried out and how it is organized to improve them.

Results
The self-appraisal criteria for the results concern what the company has achieved or is achieving. They may be expressed in the form of individual results or, preferably, as trends over a period of years. If necessary, they may be integrated and/or compared with the performance of competitors.

2.1 Customer Satisfaction
The quality approach puts customer satisfaction as the primary goal of the company. Continuous monitoring of customer satisfaction, with regard to the service rendered, should be one of the main instruments in company management. The company should describe how this is carried out, accord-

ing to customer-oriented evaluation criteria, and provide evidence of results (company's own results and possibly target competitors' results).

2.2 Employee Satisfaction

The quality approach requires that the company is able to satisfy the needs and expectations of its personnel. The company should therefore monitor personnel satisfaction continuously. The company should describe how this is carried out, according to significant evaluation criteria, and provide evidence of results.

2.3 Business Results

Business results should be viewed as the most immediate rewards for achieving the required performances and the efforts made for improving company processes. The company should describe the business results obtained in absolute values and/or trends, according to the parameters such as in the following examples.

- Financial measures

- Other result indicators such as market share, non-quality cost, and time necessary to reach the balance of the new services

2.4 Impact on Society

In the 1990s it is essential that the company achieves positive results in social terms. The approach to quality must satisfy the needs and the expectations of the community at large. In particular this includes views of the company's approach to quality of life, the environment, and to the preservation of global resources. The company should state if and how it confronts these issues and provide evidence of results.

Chapter 13:

Developing a National User-Friendly Certification and Awards Scheme

JOHN A. MURPHY

The Irish Quality Association (IQA) is a voluntary nonprofit body with a membership base of more than 2000 corporate organizations. The IQA facilitates the promotion of quality awareness through training courses, regional branch seminars, and incentive schemes such as the Quality Mark, Hygiene Mark, and award schemes.

From its inception in 1969 IQA recognized the strategic importance of quality and the need to give Ireland a new image for quality and standards. The focal point of its National Strategy for Quality first published in 1980 was to improve the competitiveness of Irish industry and services.

Today the definition of quality has evolved from its first focus on product inspection to one of TQM. The concept has grown a long way since quality was defined as "producing output in conformance to customer requirements." The evolutionary path has passed through two generations already, and leading-edge industries are pioneering third and fourth generations.

This evolution demonstrates how businesses are increasingly understanding and responding to the requirements of their customers with the ultimate objectives of increasing customer retention. This is largely due to the propensity of customers to spend more the longer their attachment to a particular company. Increased customer retention can only be achieved by an obsession with quality.

Customer Retention and Profitability

The most important reason for the emphasis on customer satisfaction is the bottom line: Good quality is profitable. Extensive research carried out by Bain & Company in the United States verified the fact that customer satisfaction is crucial.

Bain has developed a number of techniques for measuring customer satisfaction and linking it directly to corporate profitability. The simplest and most powerful gage is the customer retention rate which is defined as the percentage of customers at the beginning of the year that are still with the company at the end of the year (see Figure 13.1). In any business, the more satisfied customers are, the higher the retention rate will be. As a firm's retention rate rises, so does the longevity of the average customer account. A retention rate of 80 percent implies that, on average, customers remain loyal for five years; a retention rate of 90 percent implies a 10-year loyalty to the company (see Figure 13.2).

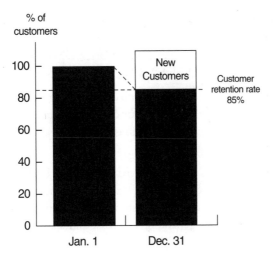

Source: Bain & Co., 1990.

Figure 13.1. Customer retention rate.

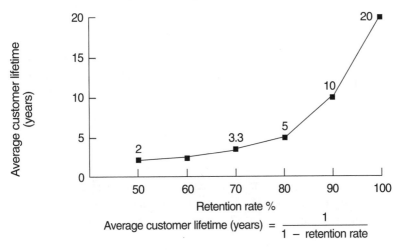

Source: Bain & Co., 1990.

Figure 13.2. Relation between customer longevity and retention.

An increase in the retention rate has a direct and dramatic impact on profit. Bain clients have experienced profitability increases of 20 percent to 125 percent from just a 5 percent increase in the client retention rate (see Figure 13.3). (The 125 percent increase was experienced by a financial services client that succeeded in raising its retention rate to 95 percent from the industry average of 90 percent.)

Long-term customers are more profitable because

- The cost of acquiring new customers can be substantial
- They tend to buy more
- They place, frequent, consistent orders and, therefore, usually cost less to service
- They are satisfied customers
- They often refer new customers at virtually no cost
- Satisfied customers are often willing to pay premium prices
- Retaining customers makes market entry or share gain difficult for competitors

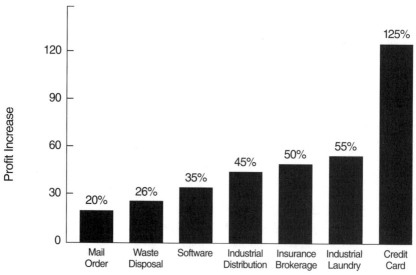

Source: Bain & Co., 1990.

Figure 13.3.   Profit impact of a 5 percent increase in retention.

The Quality Mark Scheme

In 1982 the Quality Mark Scheme was launched. It had a dual purpose.

- To encourage the application of modern quality techniques and an awareness of the importance of quality

- To give public recognition to those companies that successfully monitor and control the quality of their products or services

The initial response was encouraging with 63 firms applying immediately of which 16 were successful. Since then more than 800 firms have applied with a total of 300 companies on the register of Quality Mark holders (including 35 service companies).

The granting of the Quality Mark is based on a twofold assessment of the applicant company's QC program

- The applicant completes a detailed questionnaire which is evaluated. If the questionnaire indicates an insufficient level of QC to merit an initial audit, the applicant will be informed of these areas where deficiencies exist so that improvements may be made.

- An in-depth audit is carried out to assess the adequacy of the company's overall quality assurance program.

An application may be made by any company based in Ireland provided that the company is a member of IQA.

Some companies discover on receiving the questionnaire that their quality system is lacking to varying degrees. Most of these companies seek advice and set about implementing effective quality programs. Without participation in the Quality Mark Scheme these companies could unknowingly be heading for disaster.

A plant is inspected under eight headings, and marks are allocated for each section.

1.	Quality planning	175
2.	Incoming material control	175
3.	Manufacturing control	250

4.	Records/recording	50
5.	Environmental control	100
6.	Training	150
7.	Customer Service	50
8.	Management of product quality	150

The introduction of a special Quality Mark for service industry in 1989 marked another phase in IQA's National Strategy for Quality. There are now 35 service companies with the Quality Mark and more than 100 applications currently in the pipeline.

The audit for the service Quality Mark is based on the ISO 9004-2 guidelines and the Baldrige Award in the United States. Companies which apply for the service Quality Mark can avail themselves of an optional pre-audit consultancy. A half-day visit is arranged to discuss the requirements of an approved quality system and to assist in completion of the questionnaire.

The application procedure is similar to that of the manufacturing Quality Mark but the areas of investigation are of course somewhat different.

The auditor will explore questions in four key areas and allocate marks accordingly.

1.	Management responsibility	125
2.	Quality system	150
3.	Operational elements	500
4.	Support elements	225

In both the service and manufacturing Quality Mark Schemes successful applicants must achieve in excess of 80 percent of the total marks attainable.

The rating system operates as follows.

•	Exceptionally good	90 percent or higher
•	Good	80 percent - 89 percent
•	Marginally good	75 percent - 79 percent
•	Inadequate	50 percent - 74 percent
•	Bad	Less than 50 percent

The Quality Mark has gained consumer recognition through a series of television advertisement campaigns and through its display on product packaging. The Quality Mark is a significant marketing asset. A recent survey conducted on behalf of IQA compared reactions to a company's logo with and without the Quality Mark. This revealed that the Quality Mark can enhance perceptions of high quality service by 15 percentage points; efficiency by nine percentage points; and ability to draw new customers by 11 percentage points.

As the Quality Mark is awarded only to companies that demonstrate that their standards of quality management are rigorous and consistent, it therefore

- Represents an independent validation of the company's quality system

- Provides recognition for employees and raises morale throughout the company

- Through use of the Quality Mark on advertising, provides a quality touch-stone for customers

- Ensures that the name of the company is brought to the attention of buyers through the register of approved companies

The Quality Mark Scheme is supervised by IQA's Approvals Board. This consists of the chief executive of IQA and representatives from The Training and Employment Authority, Confederation of Irish Industries, the National Standards Authority of Ireland, the Irish Congress of Trade Unions, and Shannon Development. To ensure maximum impartiality each company is allocated a code and the identity of unsuccessful companies is not revealed to the board.

Although it is based on ISO 9000, the Quality Mark is somewhat different in its application. For instance it covers

- Plant housekeeping

- Quality costs

- Personnel appraisal

- Service quality in manufacturing

As a result of these differences, all companies applying for the Quality Mark, including those with ISO 9000, have to undergo an audit. The duration of the audit is influenced by the size and complexity of the company's operations. Most manufacturing companies on a single site location can be audited in one day which is the minimum duration audit.

The Quality Mark further differs from ISO 9000 in that

- The Quality Mark audit is a constructive audit

- The annual Quality Mark audit is a full audit and therefore appeals to companies with quality improvement programs

- The marking system associated with the audit facilitates performance measurement (benchmarking)

- The Quality Mark is more marketable than ISO 9000, particularly for consumers – they can easily identify with it

- The Quality Mark Scheme is user friendly

Benchmarking

A benchmark is a reference point against which companies can compare themselves. Leading businesses frequently use benchmarking to compare their performance in key areas against the results achieved by the best in the world. To become a high-performance organization a company must adopt world-class techniques. Only benchmarking provides information on what is required for world-class performance.

A recent study by McKinsey suggests that world-class mechanical and electronic companies may introduce new products at less than half the cost and in less than half the time it takes a typical company.[1]

The Benefits of Benchmarking

Businesses need to understand the skills and processes which are the keys to their success and to satisfying the requirements of their customers. Benchmarking can provide management with the necessary data to enable them to review critically their performance in these key areas against the best in the world. Benchmarking can produce the following benefits if performed effectively.

- Alerting management to what is possible in a world-class organization

- Setting objective performance standards for key activities to match or surpass the best in the world

- Providing insight into how other companies meet world-class standards

The marking system for the Quality Mark facilitates benchmarking and this is encouraged through the circulation of the national averages for each industrial sector and the national averages for each element of the audit in each industrial sector. For example, a Quality Mark food company will be aware of the national average for training in the food industry or the national average for quality planning and so on.

The national averages for 1992 were as follows.

- Food   88 percent

- Electronics   88 percent

- Engineering   87 percent

- Provender milling   90 percent

- Print/packaging   87 percent

- Plastics   87 percent

- Chemical   86 percent

- Healthcare   91 percent

- Service   86 percent

The National Quality Awards

A further development of the Quality Mark scheme has been the National Quality Awards. The awards were introduced by IQA to promote and encourage quality improvement in commercial, industrial, and corporate organizations. They are presented annually to companies that have demonstrated continuous improvements in quality in a given year. There are eight regional awards and a special Service Quality Award, from which one Supreme Award winner is chosen.

The selection procedure is as follows.

- All companies that have been awarded the Quality Mark are listed.

- Each award category is subdivided into industrial sectors.

- The national average for each industrial sector is reviewed.

- Companies are identified that (1) scored highest above the national average in their industrial sector, (2) obtained "Good" ratings in all areas of activity,

(3) have placed emphasis on quality planning, and (4) obtained a number of "Exceptionally Good" ratings (greater than 90 percent).

In selecting the National Quality Award winners, special consideration is given to companies that have improved on their previous year's performance.

For the Supreme Quality Award, special consideration is given to companies that actively promote quality, both internally and externally. A closer look is also taken at key areas of quality management. The number of employees is taken into account.

As in the Quality Mark Scheme the selection of award winners is made by IQA's Approvals Board. All entries are treated in the strictest confidence.

Each regional winner receives a specially commissioned trophy. A specially designed perpetual trophy is presented to the winner of the Supreme Quality Award. The awards are presented annually on World Quality Day.

Note

1. *McKinsey Quarterly*, No. 1, 1991.

Chapter 14:

Preliminary Information on the Union for Czech and Slovak Quality Awards

AGNES H. ZALUDOVA

Earlier this year a special organization, the Union for the Czech and Slovak Quality Award (UCSQA), was formed with the aim of preparing and administering the Czech and Slovak Quality Award (UCSQA). The preliminary work was done under the auspices of the Federal General Accountants Office (now disbanded) in connection with the preparation of a State Quality Policy which has not until now received government approval.

The USCQA is a voluntary, nonprofit, nongovernmental group of organizations, including the Societies for Quality, the Confederations of Industry, the Chambers of Commerce, the Unions of Employers, the Design Centres of both the Czech and the Slovak Republics, and also the consultancy organization UTRIN and the journal EKONOM. The administration of the UCSQA is located with the Czech Society for Quality (CSQ).

Currently, several proposals are under consideration on the financing, administrating, basic principles of the award, questionnaires, and methods of assessment. These proposals are based to a greater or lesser extent on the principles of the Deming Prize, the MBNQA, the European Quality Award, and the ISO 9000 series. The final version of the conditions and criteria for the award, using the most appropriate principles for the current economic situation in the country, is still under preparation.

Preparatory work is continuing despite the present political changes, the probable termination of the federation, and the changes to two separate states. The originally scheduled date of November 1993 for the first presentation of the award may have to be altered in view of contemporary events.

Chapter 15:

Report on the Argentine National Quality Award Program

Marcos E. J. Bertin

After an unusually swift process, the Argentine congress voted on a bill establishing El Premio Nacional a la Calidad. This law, which became effective September 21, 1992, was the result of efforts initiated about two years previously. FUNDECE, a private foundation for the promotion of quality and excellence created by 27 companies in 1987 and whose membership now exceeds 100, set up a committee to study the organization of a quality award in December 1990. A few months later, the president of this committee, Oscar Imbellone, had the opportunity to meet the MBNQA directors Reimann and Wayne Cassatt during a business trip to the United States. As a result, Cassatt visited Argentina in August 1991. The visit elicited the interest of a number of people and helped bring them together toward a common goal.

Earlier in 1991 one of these groups, from 3M Argentina, had taken the initiative for a national quality award into congress, seeking an interested legislator. House representative Jorge A. Lopez immediately grasped the importance of the project and organized a working team. He made this team open to anyone willing to cooperate in the making of a draft. Cassatt's visit was opportune and inspiring. Lopez's insight and openness provided the appropriate team spirit. By May 1992 the coherent contribution of many resulted in the draft submitted by Lopez and others to the House. In August 1992, FUNDECE invited Cassatt to a new visit, and at the same time, appointed Mario Mariscotti, project manager, to elaborate a proposal for the organization of the award.

The law met with very little opposition, if any. Most people regard it as a well-balanced piece of legislation: It defines the essential and is flexible with respect to organizational matters. For instance, it establishes a clear distinction between the administration of the award and the work of examiners: the former cannot intervene in the evaluation process carried out by the latter. Also, there is an independent body with the responsibility of auditing the whole process.

The public (government agencies) and private sectors are treated separately. Although from a conceptual point of view this may not be justified, it is so from a practical standpoint. For the private sector the law instructs the economy minister to call on all entities interested in the matter of quality to set a foundation. This foundation will be given the overall responsibility for the organization and management of the award program, including the evaluation process and criteria. The separation between public and private had made it possible to produce substantive progress on the private side. On December 15 more than 100 organizations, companies, and institutes met to discuss an agenda which includes the foundation by-laws, an estimate of funds needed to support the program for the first three years and the procedure to designate the first council. Two committees were formed with the purpose of studying these issues during the summer months. The work is now well within schedule and the results will be put to a vote by the assembly on March 15, 1993. That would be the foundation birthdate and the process leading to the first awards should have started.

Although the committee agreed that issues such as the organization of the foundation, the award categories, the evaluation process, and the most substantive subject, the set of criteria to be used should be left to the foundation council to decide upon and define, a few comments on the ideas exchanged about them may be of interest.

The foundation will be ruled by an administrative council of 10 people holding the overall responsibility of the whole program. Its actions will be monitored by an independent commission that will register the public reactions to the program and will make recommendations to both the council and the economy minister.

It was agreed that the council should only be concerned with policy, while an executive director would be appointed to carry out the program. Special emphasis was put on the fact that the members of the council should be persons of the highest reputation, preferably, but not exclusively, in the field of quality. To assure this, it has been proposed that the candidates for the council should be nominated in advance by the foundation members, providing evidence of the integrity and competence of the candidate.

It is also being recommended that awards should be limited to one for each of six categories involving large, medium, and small manufacturing and service companies.

The evaluation process would be carried out by an independent body of examiners and judges. As is the case in other awards, the evaluation process would consist of a first selection stage, site visits, and final selection stage. By comparison with the MBNQA, Brazilian, and Mexican systems, the possibility of reducing significantly the number of examiners is being considered. It is thought that the efficiency of their performance can be enhanced without jeopardizing the integrity and transparency of the process, by organizing intensive weeklong working sessions comprising both the individual evaluation and consensus stages.

With regard to the criteria to be used, opinions vary. Some favor adopting the MBNQA or the Mexican system. In another proposal, a set of criteria has been developed on the following basis. The purpose of the award is to promote, through public recognition of the best, the ability to compete at an international level in order to raise the standard of living of the Argentine society. The underlying postulate is that total quality is the key to achieve this. From this point of view, the award should promote the organizations fulfilling three conditions: (1) the quality of their performance results in bigger market shares, (2) they are fit to maintain the competitive edge, and (3) what they do does not affect the interests of the rest of the community. In other words: continuous success and solidarity. These conditions seem to exhaust the range of demands that society may put on the industry, while at the same time they all seem necessary.

The first condition relates to present results, including product requirements and customer and shareholder's satisfaction. The second condition refers to the ability for continuous improvement in the future and involves the organization as a whole (leadership, integrations, human resources, management, processes and measurements, and so on). And the third condition is related to the public responsibility.

A consistent set of criteria can be developed on this basis, which can be easily understood by the community, which can be universally applied, and which can provide an objective evaluation tool.

Chapter 16:

National Quality Awards: A Developing Country Perspective

Kenneth S. Stephens

Quality – A Growing Global Priority and Strategy

Quality is emerging as one of the most important disciplines and strategies for development recognized by developed and developing nations (and communities of nations). It represents a significant discipline and goal through which technical, economic, and social progress is being made, both in industrialized and industrializing countries at the national, association, corporation, institution, and enterprise levels. Quality and the quality disciplines represent a significant business – management strategy that permeates the entire spectrum of functions and organizations. This is especially true because the quality disciplines go far beyond the mere control and/or improvement of the quality of the final product or service (however important and useful that may be) to include significant reductions (improvements) in scrap; waste; repair/rework; costs; raw material, inprocess, and final product inventory; excess inspections and processing; supplier rejections; and customer complaints throughout the entire corporate or enterprise operations.

Developing countries seeking economic growth are striving to develop and expand their exports and value-added goods. This requires the improvement of the quality of locally manufactured products (and the associated processes and systems) in order to be able to compete in international markets. Likewise, the demand of the people of the developing countries for higher quality, reliability, and safety of locally produced or imported goods has been increasing.

Publicity and information from successful applications of quality methodology in the form of strategic quality management, quality systems, quality assurance, and QC applied across all functional and operational lines – as total quality – have been reaching the far corners of the globe. Developing countries eager to gain an edge are looking to these disciplines with great expectations.

Additionally, widespread adoption of the ISO 9000 series in developed and developing countries is creating further awareness of the quality disciplines. Requirements to meet ISO 9000 are being communicated to suppliers by international purchasers, even in countries where no infrastructure yet exists for assessment, certification, and registration to ISO 9000.

And while the ISO 9000 series represents a uniform consistent set of procedures, elements, and requirements that can be applied universally, it is recognized by many to lack some of the most important concepts and methodologies for achieving quality improvements and economic gains from the quality disciplines.[1]

Alternative (or additional) resources for quality management systems of total quality are the various quality awards. Among these are the Deming Prize in Japan, the MBNQA in the United States, and the European Quality Award. Other developed countries have launched national quality awards, and a growing number of developing countries are also turning to these resources in formulating, establishing, and implementing national quality awards.

National Quality Awards Survey

Motivated by an International Academy for Quality (IAQ) project, "National and International Quality Awards," together with the United Nations International Development Organization's (UNIDO's) ongoing program of technical assistance in the quality, standardization, and metrology (QSM) disciplines, a mail survey was conducted among counterparts in developing and some developed countries. The survey requested information on national quality award programs: whether any were in operation, and if so, some details as to criteria and administration; whether interest existed, and if so the purpose, expectations, objectives, national criteria, and so on; and whether there was any interest in receiving technical assistance in establishing national quality awards.

Altogether 51 countries were surveyed, 36 of the developing category and 15 of the industrialized or developed category. Definitive responses were received from 22 of the former group (61 percent) and six of the latter group (40 percent) – or 55 percent overall. A summary by country in each of the two categories follows. Included in this summary are some further details on specific programs of national quality awards reported by respondents. A review of these details is both instructive and inspiring. It reveals the extent to which these disciplines and approaches are spreading and being applied.

As the terms *developing* and *developed* are used in numerous publications with varying criteria, we define here the basis for placing a given country in the respective categories. The use of the designations is intended for convenience of comparisons and listing and does not necessarily represent or express a judgment about the stage reached by a particular country in the development process. For our purpose the per Capita GDP of 1989 is used with a threshold of $9000 (U.S. dollars) as reported in *World Statistics in Brief*.[2] Thus, countries with per capita GDP of less than $9000 are placed in the developing category for this survey.

Country-by-Country Summary: Developing Countries

Argentina

An Argentine National Quality Award was established by law 24127 in September 1992. This emerged from efforts of various groups, initially working independently, in bringing the initiative to the national congress.

It is planned to recognize the first winners of the award in October 1994 in conjunction with National Quality Month – and each October thereafter. Applications for the award will be received until mid-April 1994. The evaluation process will be carried out in the months of May through August. There will be six awards, one for each category of large, medium, and small organizations in both manufacturing and service. Argentina is seriously considering including educational and healthcare institutions in the award.

The law emphasizes that the award is to "promote continuous improvement of quality . . . in order to achieve modernization and competitiveness" and defines some of the attributes of quality such as the development, training, and participation of all members of the organization; customer satisfaction; the application of technologies for the improvement of productivity; integration with suppliers; the care of the environment; and the efficient uses of resources.

Awards for the public (for example, government agencies) and the private sectors are to be organized separately – responsibility for the latter being vested in the National Quality Award Foundation. The foundation will have the authority for the organization and administration of the award. While the law gives only an indicative list of attributes of quality, the definition of a set of specific criteria to be used in the evaluation process is vested with the foundation. The foundation will also be responsible for defining the award categories (as per the six mentioned previously).

A clear distinction is made in the law between the management of the laws and the evaluation process. The latter is to be carried out by an independent body of examiners and judges. They will be appointed by an administration council and will have autonomy with respect to the evaluation process. The process and procedures of the award, both in the public and private sectors, are to be monitored by an independent commission (overseer) which will register public reaction to the program and make recommendations to both the council and the government.

The creation of the National Quality Award Foundation for the private sector was assigned to the economy ministry. A countrywide invitation to entities interested in the matter of quality was made for a general meeting on December 15, 1992. This meeting was attended by more than 200 people from industry, commerce, academic institutions, corporate chambers, and consulting firms. A committee was established to prepare a draft of the foundation's by-laws, a budget, and a procedure to elect the first administration council. This work was carried out during the summer months (December to February) and the proposals submitted by the committee were approved in a general meeting on March 15, 1993. The foundation was thus created after 93 founding members stamped their signatures in the Act of Constitution and pledged an annual contribution of $3000 each.

The foundation is ruled by the assembly of members who delegate their authority onto the administration council. The council consists of a board of 12 members and three auditors – with a president, vice president, secretary, vice secretary, six members, and two deputies. The council will be renewed every three years by vote of the assembly. The council meets regularly on a monthly basis. However, the daily management is carried out by an executive director, appointed by the council, and assistants.

Accounting, legal advice, and other services required are contracted out to external consulting firms. The annual budget of the foundation is around $500,000 – with 60 percent coming from contributions of the members, 20 percent expected from application fees, and the remaining 20 percent from publications and conferences. Even though the law allows for allocations of the state budget, it has been decided that the program should start exclusively on private funds.

The criteria and evaluation process constitute the main core of the foundation's responsibilities. To work on these two fundamental aspects of the award an advisory committee has been set up. Awards in other countries such as the Deming Prize of Japan, the MBNQA of the United States, and the European, Mexican, Brazilian, and Colombian awards are serving as references for the make-up of the Argentine award.

The evaluation process will consist of a first selection stage, site visits, and a final selection stage. In comparison with the MBNQA, Brazilian, and Mexican systems, a significant reduction in the number of examiners is being considered. It is thought that the efficiency of their performance can be enhanced by organizing intensive weeklong working sessions, comprising both individual evaluation and consensus stages, without jeopardizing the integrity and transparency of the process.

The work of the advisory committee on the criteria and evaluation will be documented in Bases del Premio Nacional a la Calidad, which will be printed and distributed to the public well in advance of the announcement for applications for the first awards.

Brazil

A national quality award has been launched with the first winners announced at World Quality Day (November 12, 1992). The Brazilian Quality Award is based on the methodology and evaluation criteria of the MBNQA. Brazil received training and has training materials from NIST and has used the European Quality Award (via EFQM) and the British Quality Award (via BQA) as benchmarks.

Plans for 1993 include the introduction of some areas in the award to address specific national needs and to create some additional awards covering public services and educational organizations.

The Brazilian Quality Award is operated by a foundation that was created by 39 companies which have donated initially $10,000 each. In preparation for implementing the award, 152 examiners and seven judges were trained in five three-day courses in Rio de Janeiro and San Paulo.

The criteria used in implementing the award, with associated weight points, is shown in Figure 16.1.

Colombia

A Colombian National Quality Award was established in 1975 and, during the ensuing 17 years, 46 organizations were presented the award from 194 applications. A new version of the award was developed in 1991 based on the Deming Prize, ISO 9000, MBNQA, the Mexican National Quality Award, and the Colombian governmental guides. The criteria for implementing the new Colombian National Quality Award, with associated weight points, is shown in Figure 16.2.

Cyprus

Interest in a national quality award in Cyprus is keen with recognition of the principal benefit/objective of promoting quality management and quality assurance concepts to service and manufacturing industries. It is expected that the award, when developed, will be administered by a national board for quality awards that includes

Categories/Items of Evaluation	Maximum Points
1.0 Leadership	**100**
1.1 Leadership of top management	40
1.2 Value of enterprise accorded to quality	15
1.3 Management for quality	25
1.4 Team responsibility	20
2.0 Information and Analysis	**70**
2.1 Collection and management of data and information on quality	20
2.2 Comparison with competition and centers of excellence	30
2.3 Analysis of data and information about quality	20
3.0 Strategic Planning for Quality	**60**
3.1 Procedures for strategic planning for quality	35
3.2 Goals and plans for quality	25
4.0 Utilization of Human Resources	**150**
4.1 Management of human resources	20
4.2 Involvement of employees	40
4.3 Education and training in quality	40
4.4 Recognition of the efforts of employees	25
4.5 Well-being and morale of employees	25
5.0 Quality Guarantee of Products and Services	**140**
5.1 Preparation and introduction of products and services into the market	35
5.2 Control of quality of the processes	20
5.3 Continuous improvement of processes	20
5.4 Quality evaluation	15
5.5 Documentation	10
5.6 Quality of procedures in the firm and of support services	20
5.7 Supplier quality	20
6.0 Results Obtained with Respect to Quality	**180**
6.1 Results obtained with respect to quality of products and services	90
6.2 Results obtained with respect to quality in the process, in the firm, in operations, and in support services	50
6.3 Results obtained with respect to suppliers	40
7.0 Customer Satisfaction	**300**
7.1 Determination of the requirements and expectations of customers	30
7.2 Management of relations with customers	50
7.3 Patterns of service to the customers	20
7.4 Agreements with customers	15
7.5 Solution of claims with objective to improve quality	25
7.6 Determination of customer satisfaction	20
7.7 Results related to customer satisfaction	70
7.8 Comparison of customer satisfaction	70
Total Points	**1000**

Figure 16.1. Brazilian Quality Award criteria.

Areas and Criteria	Maximum Points
1.0 Customer Satisfaction	**180**
1.1 Management of customer relations	
1.2 Customer knowledge	
1.3 Feedback systems	
1.4 Results	
2.0 Leadership	**100**
2.1 Leadership by example	
2.2 Quality values	
3.0 Human Resources	**150**
3.1 Participation and involvement	
3.2 Training	
3.3 Evaluation and recognition	
3.4 Quality of life in the workplace	
4.0 Quality Strategy	**60**
4.1 Strategic planning	
4.2 Operations planning	
5.0 Quality Information	**90**
5.1 Data and sources	
5.2 Information analysis	
6.0 Quality Assurance and Improvement	**140**
6.1 Design and development of goods and services	
6.2 Control of operational process	
6.3 Control of administrative and support services	
6.4 Control of measurement and test equipment	
6.5 Continuous improvement	
6.6 Documentation and recording of quality	
6.7 Audits or evaluations of the quality assurance system	
7.0 Supplier Relations	**60**
7.1 Quality in purchases	
7.2 Suppliers and subcontractors	
8.0 Physical Plant and Installation	**60**
8.1 Installation, cleaning, and maintenance	
8.2 Industrial security and environmental control	
9.0 Effects on Environment	**60**
9.1 Preservation of the ecosystems	
9.2 Promotion of quality culture in the community	
10.0 Achievements in Improvement	**100**
10.1 Improvement of products and services	
10.2 Improvement of support areas	
10.3 Comparison of results	
Total Points	**1000**

Figure 16.2. Columbian National Quality Award criteria.

participants from the public and private sectors such as the Ministry of Commerce and Industry (that includes the Cyprus Organization for Standards), the Ministry of Finance, the Cyprus Chamber of Commerce and Industry, the Cyprus Employers and Industrialists Federation, and the Industrial Training Authority.

It is further expected that applications for the award would be published through the press with all legal Cypriot registered firms eligible to submit applications. Benefits expected by Cyprus in formulating and implementing an award include the promotion of the quality disciplines to the industrial enterprise level – with concomitant technical and economic side effects.

Following the MBNQA guidelines, Cyprus feels that the winners of its award should be required to disseminate the information through case studies that can include the company and project profile and analysis of the results to further promote applications of quality initiatives. Cyprus has expressed keen interest in obtaining technical assistance (for example, from UNIDO) to design an efficient and effective quality award scheme.

Egypt

Egypt has also expressed keen interest in formulating and implementing a national quality award. This has been conveyed by the president of the Egyptian Organization for Standardization & Quality Control (EOS), Ministry of Industry.

Remarkable changes are taking place in Egypt. Privatization of the public sector enterprises has started and there have been increasing tendencies toward achieving an open market in a few years. This has to be met by better national products, lower costs, and the adoption of high standards to be able to compete with the national, regional, and international markets.

The ISO 9000 series has been adopted as Egyptian standards and an EOS 9000 club has been organized on the same basis as the ISO 9000 forum. EOS is exerting great efforts to increase awareness of quality through workshops, seminars, and training courses based on ISO 9000. Thus, quality is being given special emphasis in Egypt and there is great interest to establish a national quality award with principal objectives of promoting and enhancing the quality disciplines at the national, society, and enterprise levels.

In support of this growing interest in the quality disciplines to help achieve an improvement in the national economy, a UNIDO project of technical assistance was carried out in Egypt. The objective of the project was to introduce TQM/QC in Egyptian enterprises and subsequently on a national scale, through a pilot top-management seminar and training/consulting workshop in a selected number of industries, that incorporate a knowledge and effective use of quality leadership, quality systems, employee involvement, quality planning, progress control, and quality management activities.

The principal counterpart for the project has been the Management Development Centre for Industry (MDCI) of the Ministry of Cabinet Affairs and Ministry of State for Administrative Development – an educational unit with excellent facilities and staff to continue the project achievements. MDCI is also cooperating with other organizations, such as EOS, in promoting these disciplines. The contractor under the UNIDO project has been David Hutching International. It is expected that this technical assistance project will serve as a catalyst for further extensions of TQM, including a national quality award.

Ghana

The Ghana Standards Bureau (GSB) has conveyed interest in instituting a national quality award, perhaps as early as 1993. Of initial interest is the manufacturing sector with extension to the service sector when implementation of applicable quality standards is made.

Principle criteria envisaged as the basis of a national quality award are (1) intent of implementation of quality systems standards, and (2) period within which a company has consistently had its products certified by GSB, according to relevant product standards.

The need for instituting the award has been felt in Ghana because it is seen as a way of promoting the use of quality principles in the Ghanian economy – both in the manufacturing and service sector. GSB would welcome any assistance, advice, and information on the experiences that other countries have had through the institution of similar awards.

India

The Rajiv Gandhi National Quality Award was launched by India in 1991. It is largely based on the MBNQA. Separate awards have been instituted for small-scale companies and large-scale companies and for each of the industrial sectors: metallurgical, electrical and electronic, chemical, food and drug, textile, engineering, and others. There is the highest award each for the small- and large-scale sectors for the company deemed the very best in the entire country. All manufacturing companies situated in India are eligible to compete.

The objectives in instituting the award are to spark involvement in quality progress, raise Indian products to high levels of quality, and better equip companies to meet the challenges of competition in the domestic and international markets. The Bureau of Indian Standards is providing all secretarial services and infrastructure for operating the award. The criteria list with point allocations is shown in Figure 16.3. In addition to these criteria, broad classification of areas in which contesting firms will be evaluated are management of quality, quality system, customer feedback system, personnel training and evaluation, quality cost control, and contribution toward environmental safety (internal and external).

Criteria	Marks Allocated
1.0 Management responsibility	100
2.0 Quality system	80
3.0 Quality in marketing	80
4.0 Design development and application	30
5.0 Quality in procurement	50
6.0 Production control, process capability, and applications of statistical techniques	80
7.0 Material control and traceability	30
8.0 Product verification	50
9.0 Control of measuring and test equipment	60
10.0 Inspection and test status	50
11.0 Control of nonconforming product(s) and corrective action	50
12.0 Handling and postproduction functions	30
13.0 After sales servicing – customer feedback system	50
14.0 Adequacy of document generation and exploitation	50
15.0 Personnel training and motivation	80
16.0 Quality cost control	80
17.0 Auditing of the quality system (internal)	50
Total	1000

Figure 16.3. Rajiv Gandhi National Quality Award criteria.

The award will

- Encourage Indian companies to improve quality for the pride and recognition while obtaining competitive edge through increased productivity

- Recognize the achievements of those companies which improve quality of their products and provide an example to others

- Establish guidelines and criteria that can be used by industries in evaluating their own quality improvement efforts

- Provide specific guidance for other organizations that wish to learn how to manage for improved quality by making available detailed information on how award-winning organizations were able to change their cultures and achieve eminence

The following are seen as benefits to award winners.

- Winning company(ies) will be reckoned as champions of the quality movement in India.

- Winners can publicize their achievements on their printed and publicity material.

- The awards will be presented in a ceremony in the presence of top industrial personalities and government officials.

- Winners will obtain leverage in business – on being recognized as the very best.

A significant quotation from the late prime minister Shri Rajiv Gandhi that serves as a basis for naming the award in his honor is

> The need of the hour is a national commitment to quality in all walks of life. We should not be satisfied with anything but the best in the goods and services that we produce.

Indonesia

While no national quality award has as yet been established, Indonesia, through the Dewan Standardisasi Nasional (DSN), has a plan to establish such an award. Officials in Indonesia are presently studying the criteria, the structure, and the management of the national quality awards being administered in other countries. They would also like to have assistance (for example, through UNIDO) in formulating, establishing, and implementing a national quality award.

Jamaica

The standards council (overseeing the Jamaica Bureau of Standards) and government officials have expressed keen interest in the establishment of a national quality award in Jamaica. Expected participants and beneficiaries include members of the Jamaica Manufacturers' Association, service organizations, and others, and in particular, the certified manufacturers and adherents to the ISO 9000 series.

Jamaica foresees the objectives of the award to challenge Jamaican producers and service companies to strive for excellence as well as to recognize and publicize successful quality strategies and efforts. It further foresees an award ceremony presided over by the governor general – with the award managed by a committee of personalities dedicated to quality processes and quality initiatives.

The government's emphasis for national growth is on production and export – with quality seen as a platform for productivity. Hence, it is seen that the award would be felt from a national perspective.

The Jamaica Bureau of Standards welcomes technical assistance for establishing its national quality award.

Kenya

No national quality award yet exists in Kenya and no decision has been taken to start one. However, as reported by the Kenya Bureau of Standards, an advanced product certification system exists. Manufacturers that meet the requirements of Kenya standards are licensed to display a mark of quality on their products. It is expected that the product certification system should eventually become the basis of a national quality award.

This issue of quality is beginning to emerge in Kenya, partly as a result of the emphasis on exporting. In March 1990 a subcontracting seminar was held in Nyeri, jointly organized by United Nations Development Program (UNDP), UNIDO, the International Labor Organization (ILO), and the Ministry of Industry. During the seminar, it emerged that it would be difficult to establish a program on subcontracting at enterprise level unless the quality of goods and services from the enterprises was guaranteed. It became important to look into ways of raising concern for quality, especially at the enterprise level.

Thus, in August 1990 a roundtable discussion meeting was organized in Nairobi by UNDP, UNIDO, and the Ministry of Industry. The title, "Raising the Concern for Quality in Kenya – Towards an Action Agenda" was given to the meeting. Participants at the roundtable were drawn from government ministries, research institutions, trade unions, and representatives of manufacturers.

Emerging from this roundtable was the formulation and subsequent implementation of a UNDP/UNIDO technical assistance project entitled, "Total Quality Management in Industrial Enterprises." The principal counterpart for this project is the Kenyan Federation of Employers (KFE). Under the project, three national experts have been recruited to understudy the principal subcontractor, Juran Institute, and assist in implementing TQM in as many as 15 model enterprises. The project includes top-management seminars, facilitator workshops and training programs, study tours and fellowships abroad, and equipment for training and promotion in the quality disciplines. It is expected that this program will contribute to the establishment of a national quality award.

Korea

The Republic of Korea is implementing an entire series of quality-related awards: (1) national QC awards, (2) production innovation awards, (3) value engineering (VE) awards, (4) total productive maintenance awards, (5) outstanding Korean Standards (KS) marking factory awards, (6) research team awards, (7) QC circle awards, (8) literature awards [QC, industrial engineering (IE), VE, and TPM fields], and (9) outstanding individual awards.

The national QC awards consist of two awards (Grand Prize and Excellence Award) in two categories (large-scale industry and small- and medium-scale industry). The documentation required for application includes the application form, 25 copies of a QC activities description, 25 copies of a QC improvement case study book, and a description of consulting and training by outside consultants in the last year.

The award is administered through the QC Department, QC Bureau, of the Industrial Advancement Administration (IAA). A document assessment is carried out to determine whether a follow-up field visit will be made. When warranted the field visit is conducted by eight experts from academic and relevant disciplines. This usually consists of a three-day period but may be extended in the case of multiple factories. The applicant is expected to prepare written materials that confirm descriptions submitted with the application – and each process or relevant department has to present daily control materials to verify the system.

Assessments are carried out according to eight categories or criteria as outlined in Figure 16.4, with point weights. The assessors examine how well each observing point has been performed and make classifications in five degrees or states as follows: (1) A: very excellent, (2) B: excellent, (3) C: average, (4) D: insufficient, and (5) E: not performed. The reports of the assessors are then examined by an Award Judging Board. The board consists of 12 persons with the deputy administrator of IAA as chairman, the director general of the QC Bureau of IAA as vice chairman, and other members made up of assessors and representatives of the Korean Standards Association (KSA) and the Korean Society for Quality Control.

Scoring for the awards is as follows

1. For the Grand Prize
   - More than 450 points for large companies or more than 425 points for small and medium companies
   - A "B" or higher degree

2. For the Excellence Award
   - More than 425 points for large companies or more than 400 points for small and medium companies
   - A "C" or higher degree

The awards are presented at the National Quality Control – Standardization Convention.

Benefits to the awardees include, (1) they can publicize their award, (2) when the factory has earned the factory degree mark or KS marks (for products), it can be exempted from the follow-up assessment for longer periods, and (3) voluntary public presentations are held within six months of receiving the award, actively cooperating with IAA and KSA in promotion case presentation, seminars, and project consulting.

Malaysia

Malaysia has been implementing a national quality award since 1990 – referred to as the Prime Minister's Quality Award. In addition there are awards by the Ministry of International Trade and Industry (MITI) for product excellence, export achievement, hotels, and so on. The Standards and Industrial Research Institute of Malaysia (SIRIM) – administrator of the quality award – also gives out awards for manufacturing excellence to ISO 9000 and for products conforming to Malaysian standards.

Category	Items	Maximum Points
1. Policy Plan		
	Company Policy and Plan **50**	
Observing Points:	1. Enthusiasm and awareness of top management	
	2. Policy and plan on management and QC	
	3. Establishment of policy and reasonableness of its deployment	
	4. Employees' understanding and awareness of the policy	
	5. Appropriateness of control items for the achievement of the policy and implementation situation of the items	
	Effects and Future Plan **20**	
Observing Points:	1. Management results and anticipating effects in terms of quality, quantity, and cost	
	2. Quality comparison with domestic and overseas products and change of market share	
	3. Identification of problems, measures, and future plans	
2. Organization, Training, and Standardization		
	Operation of Organization **20**	
Observing Points:	1. Reasonableness of job allotment and empowerment	
	2. Cooperation between departments and coordination systems	
	Education and Training **20**	
Observing Points:	1. Education/training systems, execution, and assessment	
	2. Training of divisional QC experts and utilizing results	
	Standardization **30**	
Observing Points:	1. Promotion policy and system of standardization	
	2. Establishment of standards, training, and its observance	
	3. Operation results of proposal system and promotion for its activation	
3. Information Control and Computerization		
	Information Control **30**	
Observing Points:	1. Collection, analysis, and utilization system of information	
	2. Accumulation, utilization, and patent applications of technical information	
	3. Operation and activation of proposal system	
	Computerization **40**	
Observing Points:	1. Policy and plan for computerization	
	2. Development of computer package program	
	3. Utilization of computerized materials by executives and managers	

Figure 16.4.   Korean national quality control award assessment criteria.

4. Activities of Quality Circle and Morale Enhancement

Activities of Quality Circle **30**

Observing Points:
1. Promotion policy and operation system of quality circles' activities
2. Results and methods of activation of quality circles' activities
3. Appropriateness of education for QCC members
4. Supporting system (incentives) for QCC activities

Morale Enhancement **30**

Observing Points:
1. Morale enhancement system for employees
2. Welfare system
3. Working conditions and safety control
4. Survey of employee's morale and its utilization (analysis of turnover of employees, handling of complaints)

5. Development of New Products

Development of New Products **60**

Observing Points:
1. Policy, system, and results of development of new products (results of the latest three years)
2. Identification of consumers' needs and reflection in the design quality
3. Determination of design quality and assessment method on trial and trial products
4. Execution, assessment, and assurance method of reliability test
5. Early stage of flow control system and operation of developed products

6. Manufacturing Process Control

Manufacturing Process Control **60**

Observing Points:
1. QC process chart, check points in processes, and clarification of control methods
2. Control chart and identification and utilization of process capability
3. Measures for reduction of defective processes and clear actions when out of control
4. Improvement activities to solve major quality problems
5. Education and training of workers and observance of operation standards

7. Quality Assessment, Control of Facilities, and Measuring Instruments

Quality Assessment and Inspection **20**

Observing Points:
1. Appropriateness of test and inspection methods and clarification of their judging standards
2. Handling and prevention measures on defective lots and products
3. Education and training and function of test inspection personnel

Figure 16.4. *(continued).*

7. Quality Assessment, Control of Facilities, and Measuring Instruments

	Quality Assessment and Inspection **20**
Observing Points:	1. Appropriateness of test and inspection methods and clarification of their judging standards
	2. Handling and prevention measures on defective lots and products
	3. Education and training and function of test inspection personnel

	Control of Facilities and Measuring Instruments 30
Observing Points:	1. Productive maintenance methods and security and operations of components supplied
	2. Possession of measuring instruments, maintenance of their precision, and personnel management
	3. Identification of capability of facilities, analysis and utilization of failure statistics

8. Purchasing, Subcontracting, and Sales Control

	Purchasing, Materials, and Subcontracting Control **30**
Observing Points:	1. Supplier selecting standards and inspection methods of products purchased
	2. Control of optimum inventory and keeping of inventory quality
	3. Subcontractor selecting standards and operation and assessment methods

	Marketing, Sales, and Handling of Complaints .. **40**
Observing Points:	1. Survey on consumer satisfaction on the applicant's quality and service and its utilization
	2. Handling system of consumers' complaints and prevention measures
	3. Survey on consumers' complaints and utilization of related information

TOTAL **500**

Figure 16.4. *(continued).*

In line with the Vision 2020 aspirations, the government has formulated a National Development Policy that aims to achieve a balanced, broadbased, and international competitive economy. The industrial sector is expected to continue to spearhead the growth of the economy to achieve this objective. The development of the service sector will also be given priority to enable it to assume a leading role as a major contributor to the growth of the economy.

With the presentation of prestigious awards Malaysia hopes to inculcate excellence in all areas to achieve these national goals. It is expected that everyone will strive to be competitive and continually enhance their efforts to increase quality and productivity so that the nation will become known as a reliable supplier of high-quality goods and services in the international marketplace.

The Prime Minister's Quality Award has the following purposes.

1. To encourage and promote quality awareness

2. To give formal recognition to agencies that have in-depth understanding of quality improvement and management and also to agencies that have attained preeminent quality leadership

3. To permit agencies to publicize and advertise receipt of the award and to share their successful quality strategies

4. To encourage healthy competition among respective agencies toward the betterment of quality management practice

The recipient of the award receives a cash prize of M $30,000, a trophy, and a certificate of recognition. Agencies are allowed to use the symbol "Q" for the period of three years and to use the statement, "Winner of Prime Minister's Quality Award 199 –," printed at the head or foot of their letterhead.

The basic eligibility for the award is to public or private agencies incorporated and located in Malaysia. The system for scoring is based on three evaluation dimensions: (1) approaches introduced, (2) deployment of the approaches, and (3) results obtained through the application of the approaches. For this latter dimension the performance indicators used to assess results are: (1) reduction in operational cost, (2) increase in output, (3) time saved in performing particular jobs, and (4) increased customer satisfaction. The award examination addresses all aspects of a TQM system and quality improvement results through eight categories or criteria, namely leadership, information and analysis, strategic resource utilization, quality assurance of output, quality achievements, customer satisfaction, and outstanding innovations. Winners of the award are divided into four categories with the best candidate receiving the Prime Minister's Quality Award; the runner-up receives the Chief Secretary Award; and the third and fourth runners-up receive the Public Service Chief Director Award and Manpower and Modernization Planning Unit Chief Director Award, respectively.

The Industrial Excellence Awards consist of four awards: the Product Excellence Award, the Export Excellence Award, the Quality Management Award, and the Hotel Performance Award. The Quality Management Award, in particular, is a development from the National QCC Award that was instituted in 1985 by the National Productivity Centre for quality management practices. It is open to any organization registered under the Companies Act in Malaysia. It is based on the following criteria.

- Demonstration that its management has a good understanding of the requirements for quality management

- Evidence of total commitment of top management

- Active participation by all levels of employees in problem solving and quality improvement activities

SIRIM's awards include the Manufacturing Excellence Award and the Product Certification Marking Scheme. The former is given to industries through the quality system Certification Scheme that certifies that the quality system operated by the company meets the requirements of ISO 9000. The latter award is granted if the product complies with the applicable standard and the factory has a QC system that meets the requirements of SIRIM's Scheme of Supervision and Control.

Mexico

A national quality award was established in January 26, 1988, as published in the Official Diary of the Federation, articles 80 and 81 of the Federal Law on Metrology and Standardization. The award is divided into six categories: (1) large industrial enterprises, (2) small and medium industrial enterprises, (3) large commercial enterprises, (4) small and medium commercial enterprises, (5) large service enterprises, and (6) small and medium service enterprises.

The principal objectives of the award are (1) to promote and stimulate the establishment of procedures for total quality in the productive units of goods and services in Mexico, (2) to promote higher productivity in the various economic activities by increasing the efficiency of the productive processes and the quality of the products from the point of view of promotion and not of regulation, and (3) to promote the exports of national products, goods, and services based on better quality and thus obtain a larger level of competitiveness and prestige in the international markets.

For enterprises to apply they must (1) have a sustained process of total quality in their areas of production of goods and services as well as in their administration and distribution areas, (2) have systems and processes to achieve total quality as well as qualitative and quantitative documentation that the results have been achieved, and (3) prove that the enterprise has not received any sanction on the part of the Secretary of Commerce and Industrial Promotion in the immediately preceding year from the publication of the award announcement.

A preliminary evaluation is carried out based on an initial questionnaire completed by the applicant. The questionnaire focuses on customer satisfaction, leadership, human resources, information and analysis, planning, quality assurance, environment, achievements, and quality philosophy.

The second phase evaluation for the finalists (enterprises) is based on three aspects: (1) focus or strategy, (2) implantation or implementation, and (3) results or achievements. The categories and specific terms of the criteria for the final evaluation, with maximum point values, are shown in Figure 16.5.

Nigeria

The Standards Organization of Nigeria (SON) has a national quality award under its certification marking programs. The award is based essentially on the organization's requirements for the Nigerian Industrial Standards (NIS) certification mark. The NIS certification marks of quality awards are given annually on October 14 as part of a series of programs to mark World Standards Day.

The award was introduced in 1974 with Lever Brothers Nigeria PLC as the first winner. Since then more than 100 manufacturing companies have qualified for the award.

A recent innovation to this type of award has been made whereby companies that have won the award consistently for more than 10 years are given gold certificates and gold plaques, while those that have won the award for a period of from five to nine years consistently are given silver certificates and plaques. Those companies that have won the award for a year up to four years consistently are given ordinary certificates only.

Winners whose products fail to meet the quality NIS certification mark in any given year have their certificates withdrawn and they are statutorily prevented from using the NIS certification mark logo on their products. In other words, the

Categories and Specific Terms	Maximum Points
1.0 Customer Satisfaction	**180**
1.1 Customer knowledge	60
1.2 Feedback systems	60
1.3 Standards of service	30
1.4 Results	30
2.0 Leadership	**100**
2.1 Leadership by example	70
2.2 Quality values	30
3.0 Human Resources	**150**
3.1 Involvement	40
3.2 Training	50
3.3 Recognition	30
3.4 Quality of life in the workplace	30
4.0 Information and Analysis	**100**
4.1 Data and sources	70
4.2 Information analysis	30
5.0 Planning	**80**
5.1 Strategic planning	30
5.2 Operations planning	50
6.0 Quality Assurance	**160**
6.1 Design and control	30
6.2 Continuous improvement	70
6.3 Suppliers	30
6.4 Documentation	30
7.0 Effects on Environment	**80**
7.1 Preservation of ecosystems	30
7.2 Development of small and medium suppliers	50
8.0 Results	**150**
8.1 Improvement of products and services	60
8.2 Improvement of support areas and suppliers	40
8.3 Comparison of results	50
Total Points	**1000**

Figure 16.5.   Mexican National Quality Award criteria.

certificates or awards are valid for one year only and are renewable yearly. Classification of the awards into ordinary, silver, and gold certificates/plaques has brought very keen competition among the various manufacturing companies within the country and they all strive vigorously to retain and even improve the quality of their products in order not to lose the award.

The benefits accruing to the award winners have been tremendous in recent times in the sense that government and nongovernment bulk purchasers now make the winning of the awards and the NIS certification mark a condition for contract patronage of their products. The scheme has really popularized both the activities of the SON and the products of the award winners, hence the number of manufacturing companies that qualifies for the award increases every year. SON also prints annually a directory of NIS-certified quality products for the use of buyers and the general public.

There is no other organization that gives a quality award on a national scale as yet. Plans are under way for the Quality Control Society of Nigeria that was formed in 1990 to introduce a quality award program. Nigeria has expressed interest in receiving technical assistance (for example, from UNIDO) in the establishment of a national quality award.

Pakistan

No national quality award is presently administered in Pakistan. However, the government of Pakistan is instituting awards on highest overall exports, highest engineering goods exports, best export publicity, highest unit values in textiles, and introduction of entirely new export products.

According to the Pakistan Standards Institution, Ministry of Industries, a national quality award is now overdue and should be introduced on a priority basis. It has expressed keen interest in obtaining technical assistance (for example, from UNIDO) to formulate, establish, and implement an efficient and effective national quality award.

Philippines

At present the Philippine Society for Quality Control and the Philippine Productivity Movement, in cooperation with the Board of Investments and the Bureau of Product Standards, Department of Trade and Industry, are conferring annual awards to outstanding quality companies. This is referred to as the Outstanding Quality Company of the Year (OQCY) Award. The objective of the award is to stimulate interest in the advancement and dissemination of quality management practices in the industry, service, and other areas by giving due recognition to the companies that have shown outstanding achievements through exemplary quality improvements. Any company or corporation that is doing business in the

Philippines and is properly registered under the Philippine laws may be nominated for the award. Winners are selected by a group of distinguished judges, whose decisions are final and nonappealable. The winners are selected on the basis of their achievements or contributions in the following areas.

1. *Institutionalizing TQM Processes/Systems.* This category examines the company's development, promotion, dissemination, and improvement programs on the use and applications of quality management practices. It is assigned a total of 385 points or 38.5 percent of the award criteria.

2. *Quality Results.* This category examines the company's performances in the improvement of quality and reliability of the company's products and/or services. It is assigned a total of 200 points or 20 percent of the award criteria.

3. *Assistance Rendered to Vendors and Other Entities.* This category examines the systematic approaches used by the company to achieve TQM of products and services including assurance of quality of procured materials, parts, and services. It is assigned a total of 40 points or 4 percent of the award criteria.

4. *Outstanding Contributions to Consumer Satisfaction.* This category examines the company's processes, selection of measurable standards, and so on, used to evaluate and improve methods for determining customer satisfaction. It is assigned a total of 200 points or 20 percent of the award criteria.

5. *Increased Profitability as a Result of Managing Cost of Quality (COQ).* This category examines improvement trends in COQ for at least the past three years. It is assigned a total of 175 points of 17.5 percent of the award criteria.

A more detailed listing of the criteria for implementing the Philippines OQCY Award, with associated weight points, is shown in Figure 16.6.

The evaluation process includes: receipt of applications; first-stage review by six to eight examiners; a decision whether the applicant receives a site visit, made by judges based on the initial review, with a feedback report sent to all applicants not selected for a site visit; conduct of an extensive site visit by four to six examiners; and recommendation of the winners by judges, with a feedback report sent to all applicants not recommended.

From 1987 to 1991 six companies have been awarded the OQCY Award: 1987 – Avon Manufacturing Inc.; 1988 – Motorola Phils. and Johnson & Johnson Phils.; 1989 – Intel Manufacturing Phils.; 1990 – Texas Instruments (Phils.); and 1991 – Electronics Assemblies.

Interest exists in expanding the program for improved benefit to the trade and industry sector – that might involve information and promotion programs as well as conduct of assessment of competing companies. The Philippines has expressed keen interest in receiving technical assistance in expanding and improving its national quality award.

Poland

No national quality award, like the Deming Prize in Japan, the MBNQA in the United States, and the European Quality Award, has as yet been established in Poland. Instead, at least in the past, various kinds of national quality competitions were organized in Poland, such as the All-Polish Competition of Good Work, DO-RO in the 1970s, and the All-Polish Competition Quality-89. These recognitions have resulted in implementation of zero-defect methods of work and operator control and improvement in organization and quality production management. Prize winners/enterprises of these competitions were given cash prizes and certificates.

However, Poland has found that the organization of these competitions was quite a costly undertaking which was one of the main reasons why the activities have been discontinued. At the moment, despite the great need and interest in a national quality award that would play an important role in promoting and enhancing the quality of products and services at the national, social, and enterprise levels, it is not feasible to establish and implement such an award due to financial deficiencies. It is hoped that, as the social-economic situation in Poland improves, it will be possible to formulate, establish, and implement a national quality award. Poland has expressed interest in technical assistance and support directed toward the establishment of such an award.

Saudi Arabia

The Saudi Arabian Standards Organization (SASO) is applying regulations for licensing national firms to use the quality mark on commodities produced that conform to Saudi Arabian standards. Through the program, SASO encourages national

Categories and Specific Terms	Maximum Points
1.0 Institutionalizing TQM Processes/Systems	**385**
1.1 Top management leadership and commitment	100
1.2 Strategic quality planning	60
1.3 Continuous education and training	40
1.4 Quality network for employee involvement in problem solving and decision making	40
1.5 Employees recognition award	25
1.6 Open communications	25
1.7 Documentation	25
1.8 Process design control	35
1.9 Active and effective corrective actions	25
1.10 Previous national or international recognitions received as a result of pursuing good practices in quality management	10
2.0 Quality Results	**200**
2.1 Quality of products and services	90
2.2 Operational and business process quality improvement	55
2.3 Quality improvement application	55
3.0 Assistance Rendered to Vendors and Other Entities	**40**
3.1 Effectiveness in linkage/partnership between the vendee and vendor	20
3.2 Development and assistance given by company to its suppliers	20
4.0 Outstanding Contributions to Consumer Satisfaction	**200**
4.1 Steps taken in getting to know more about external customer's requirements	50
4.2 Steps taken to conform to customer needs and expectations	50
4.3 Steps taken in setting customer service standards	50
4.4 Steps taken to benchmark products and services against competitors and/or best in class	50
5.0 Increased Profitability as a Result of Managing COQ	**175**
5.1 Prevention cost	35
5.2 Appraisal cost	35
5.3 Internal failure cost	35
5.4 External failure cost	35
5.5 COQ as a percentage of net sales or as a percentage of goods sold	35
Total Points	**1000**

Figure 16.6.   Phillipine OQCY Award criteria.

firms to produce quality products – and believes that a national quality award is one of the ways for achieving this goal. Hence, SASO is keenly interested in the provision of technical assistance for the formulation, establishment, and implementation of a national quality award.

South Africa

While no national quality award now exists, the South African Bureau of Standards (SABS) has concluded that the advantages outweigh the disadvantages, if administered properly. Hence, the bureau is presently studying the relative merits of introducing a national quality award. To this end it has entered into discussions with EFQM, the Department of Trade and Industry in the U.K., the Institute of Quality Assurance, BQA, and the MBNQA administrator. The bureau has expressed interest in the conclusions and information obtained through this study.

Tanzania

The Tanzania Bureau of Standards is interested in launching a national quality award also. Hence, it is interested in receiving technical assistance to initiate a national quality award that it considers essential as a means of motivating quality progress in Tanzania.

Thailand

In Thailand, awards have been administered by the Ministry of Industry since 1981. The Thai Industrial Standards Institute (TISI) participates in the selection of distinguished manufacturers of the year. The awards are presented on the occasion of the anniversary of the Minister of Industry each year and are based on the following criteria: background and establishment, organizational structure and administration, management and production, safety at work, energy consumption, environmental management, and social activities and responsibility.

Trinidad and Tobago

No quality award has as yet been established in Trinidad and Tobago. However, the Trinidad and Tobago Bureau of Standards has been observing the examples of Japan, South Korea, the United Kingdom, and the United States with respect to governmental policy positions taken with respect to advocating quality as a means of improving industrial competitiveness.

In 1991 the government of Trinidad and Tobago began discussions with the World Bank with a view to sourcing funds for a Business Expansion and Industrial Restructuring (BEIR) project. One part of that project is a quality component which will be administered by the Trinidad and Tobago Bureau of Standards. That part of the project includes, as a major component, the establishment of a national quality system (NQS).

The objectives of the NQA would be to develop an awareness within industry and government of international criteria (ISO 9000) for effective quality management systems, establish a national system for registering/certifying enterprises which conform to these criteria, gain international recognition for this Trinidad and Tobago quality registration system, and introduce the Trinidad and Tobago industry to the benefits of TQM.

A quality awareness campaign is a major aspect of the quality component and with it is expected the establishment of a national quality award. Funds from the BEIR loan are expected to be available by the end of 1992 and a national quality award in place before the end of 1993.

Country-by-Country Summary: Developed (Industrialized) Countries

Canada

No national quality award, comparable to the MBNQA, exists in Canada. However, Canada does single out a number of Canadian businesses each year for its Canada Award for Business Excellence. The 1992 *Guide to Federal Programs and Services* describes the program as follows.

> These annual awards honor outstanding business achievers in eight categories: invention, innovation, industrial design, entrepreneurship, environment, marketing, small business, and quality. They are open to businesses of all sizes and in all fields of economic activity. Independent private sector experts select up to three awards in each category.

Israel

The Association of Electronic Industries (AEI) has established the Israeli Quality Award. The first winners were awarded in 1990 when National Semiconductor and Motorola (Israeli manufacturing subsidiaries) were selected. In 1991 AEI split the award into two categories – one for large companies (employing more than 200 persons) and one for small companies. Elbit Defense Systems received the award in 1991.

The award is similar to the MBNQA. It is open to the 90 companies that make up Israel's electronics industry. An average of 67 percent of these companies' output is exported. The award is based on stringent criteria spanning the full range of company operations, including leadership for quality, quality planning and implementation, human resource development, customer satisfaction, continuous improvement, quality infrastructure, product and service quality, and information system quality.

In support of the award, AEI has published a booklet on TQM. The Standards Institution of Israel (Quality & Certification Division) is planning a national quality award for 1993 based on similar criteria to the AEI award.

New Zealand

In response to calls from business for a national quality award in New Zealand, based on the success of the MBNQA that also served as a model for the Australian Quality Awards, the minister of commerce agreed to establish a Steering Committee with a view to exploring the potential for a New Zealand National Quality Award (NZNQA).

The Steering Committee was chaired by Doug Matheson. His background is as director of the Total Quality Management Institute, former interim Chief Executive of the Wellington City Council, and retired IBM executive and board member. Other members of the steering committee were Interlock Industries, New Zealand Manufacturers' Federation, Telecom, Toyota, New Zealand Trade Development Board, and the Ministry of Commerce.

Following consultation with the business sector and an examination of overseas models, the committee determined that the NZNQA should be a premier award with trans-Tasman and international standing, and should acknowledge excellence in the management of quality and provide participating firms with a benchmark against which they could measure their approach to quality. The committee agreed that the award should be led by the private sector and have the personal involvement and commitment of senior business leaders, as well as the support of government. To this end, the committee proposed the formation of a privately funded Quality Award Foundation.

Senior business people from leading New Zealand companies which are committed to quality were personally canvassed to obtain an indication of their support for the award as proposed by the Steering Committee. Strong support for the concept was received and the level of interest was such that the committee was encouraged to bring these parties together with a view to forming a Quality Award Foundation.

This meeting was held in May 1992 and there was general agreement among those present for the establishment of a premier NZNQA as proposed by the Steering Committee. The members at the meeting also agreed to the concept of a Quality Award Foundation and that membership of the foundation would not be limited to those present at the meeting. The meeting was also attended by the deputy prime minister and the minister of business development.

The purpose of the New Zealand Quality Award Foundation is to lead the quality awards and to generate funding and support. The foundation was established in October 1992 and is coordinating the first award to be granted in 1993. For the first year the 1992 MBNQA criteria are being used.

There are three main bodies that have an influence on quality in New Zealand.

1. Total Quality Management Institute that promotes the awareness, understanding, and education in total quality principles and practices to chief executives and senior managers. It is a corporate membership funded from subscriptions and some trading activities and operates an annual conference.

2. New Zealand Organization for Quality (NZOQ) that has been in New Zealand for many years operating a network of regional groups that focus on quality at the middle management and practitioner level. It has strong influences on the ISO 9000 area and run many education and information activities. It also runs an annual conference and operates independently of other organizations.

3. New Zealand National Quality Award Foundation.

There are two other quality awards operating in New Zealand. The Railfreight Award for Excellence in Manufacturing has been active for three years. It uses the MBNQA criteria but focuses on manufacturing only and can evaluate a single division of a larger organization. There is one winner each year. Another award is a Business Development Quality Award operated by the government-sponsored Business Development network. These are intended as encouragement awards for organizations starting the quality journey. There is one award per region (21) and a national winner from those. The criteria are the same as the NZNQA and there is a minimum and maximum points range. The organizations over the maximum are encouraged to apply for the NZNQA. A third New Zealand Quality Prize for an individual quality improvement project has been in place for a number of years. It is sponsored by the NZOQ and TELARC, the national accreditation agency.

Singapore

Since January 1992, the Singapore Institute of Standards and Industrial Research (SISIR) has been aggressively promoting the ISO 9000 series standards with local companies. SISIR considers that, at this point in time, its efforts should be focused on expanding and assisting local industries to adopt and meet the ISO 9000 series standards.

It has been exploring the possibilities of launching a national quality award, based on the criteria of the MBNQA, and is keenly interested in the findings with other developing countries and in information on technical assistance for formulating and implementing a national quality award.

Sweden

Sweden launched a Swedish Quality Award in 1992 that resembles the MBNQA. Seven criteria have been chosen for the award: leadership; information and communication; strategic quality planning; staff engagement, development, and participation; quality in working processes; results; and customer satisfaction. Integrated within these criteria are the requirements of the ISO 9000 series. Further information is available from the Swedish Institute for Quality.

United Kingdom

A quality award presently exists in the United Kingdom that is similar to the MBNQA that is administered by the BQA. However, it is still quite a topical subject in the United Kingdom since the Department of Trade and Industry has recently set up a special committee made up mainly of senior industrialists to study the possibility of establishing a new prestige award for quality for British business, building on the existing British Quality Award.

In Conclusion

Where national quality awards have not as yet been established, there appears to be universal interest in obtaining information and technical assistance in formulating, establishing, and implementing such awards. A considerable variation in types of awards is evident – with some countries basing their awards primarily on their product quality marking and certification programs and newly developed quality systems certification to ISO 9000. Others combine with quality, export promotions, product innovation and development, business excellence, safety and environmental protection, and so on. However, most of these recognize the distinction and value of a broad-based national quality award that is similar to the popular Deming Prize, MBNQA, and European Quality Award.

The national and international communities of quality professionals and related organizations have a serious responsibility (with related opportunities) to provide promotion and technical assistance to developing countries (in particular) to establish the infrastructure for implementing total quality. A very viable mechanism is in the form of a national quality award that includes essential features and, at the same time, recognizes unique national characteristics that will be beneficial in enhancing quality and, through the quality disciplines, improving the economy and environment.

Notes

1. The author has retired recently from the United Nations International Development Organization (UNIDO) where he was unit coordinator of the quality, standardization, and methodology (QSM) program. The views expressed are those of the author and do not necessarily represent those of UNIDO. Kenneth S. Stephens, Quality Systems & Certification – Some Observations and Thoughts, Transactions of the Hungarian Quality Week Conference, Budapest, Hungary, September 13-18, 1992, and *EOQ Quality*, No. 1, March 1993, pp. 5-12.

2. *World Statistics in Brief*, United Nations Statistical pocketbook, 14th ed. Department of Economic and Social Development, Statistical Office, Series V, No. 14, United Nations, New York, 1992.

Chapter 17:

A Critical Review of Methods for Quality Awards and Self-Assessment

TITO CONTI

Today, it would be limiting to consider quality awards solely in relation to the actual award – what to reward, what criteria to use, how to ensure the effectiveness of the assessment process. Experience has shown – specifically, the MBNQA – that, for every company competing for an award, 1000 other companies use the award model and assessment criteria to conduct self-assessments and diagnoses, to take stock, and to log their positions on the long hard road toward total quality. Both possible applications of the TQM model and assessment criteria should therefore be borne in mind when a new award is planned or improvements are made to an existing award: identification and public recognition of excellent companies on one hand, self-assessment for diagnostic and improvement purposes on the other.

But this alone is not enough today. A clear distinction must be made with reference to TQM models. With reference to assessment methods, according to whether the intention is to commend excellence or to encourage self-assessment, two separate paths should be followed after a certain point. It is fairly clear that the model is influenced by an overall objective. The sole purpose of self-assessment is company improvement, while awards usually aim to recognize excellent companies. But the emphasis can be placed on different aspects of excellence of the quality system, and so on. There is still ample leeway for the exercise of discretionary powers in this field, as the different choices made by the different award bodies show. This review is therefore intended as a contribution to a debate launched in 1990 by a paper prepared for the European Quality Award, which has adopted some of the principles recommended in that paper, in particular, the central role of self-assessment as the basis for all external assessments, especially award-oriented assessments.

If an award is meant to encourage self-assessment, then self-assessment should take absolute precedence when the TQM model is defined. Otherwise, it is best to keep the models for self-assessment and for awards separate. The self-assessment *method,* that is, the way in which the self-assessment is conducted, should be defined separately. It is not legitimate simply to extrapolate the methods used by award-oriented assessments (for example, the MBNQA's approach-deployment-results method.) Readers will point out that these distinctions have not been observed in the past. This is so. The TQM models developed for quality awards have automatically become self-assessment models, as have the assessment methods. This is not a flaw in the awards, but exactly the opposite. One of the merits of the awards – particularly of the MBNQA – is that they have increased awareness of the importance of self-assessment in relation to a precise TQM model, performed according to precisely defined assessment methods. This was a case of "the tail wagging the dog."[1] With today's greater awareness of self-assessment as a variable in its own right, it is important that the dog should regain its proper role; that self-assessment should be seen as the cornerstone of the continuous improvement process and that external assessments such as award-oriented assessments, while pursuing their own particular objectives, should largely be based on company self-assessments.

If this reasoning is correct, the interests of clarity would be served if awards were kept somewhat separate from self-assessment. Today's quality awards tend to be associated with a particular nation or region, and this encourages the development of nationalist attitudes dogmatically focused on the respective TQM model. The danger where self-assessment is concerned is that the American – or European or Japanese – model will be taken as the basis of reference, instead of a model molded by experience and tailored to the specific characteristics of the company and the market in which it operates.

It would be a disaster for quality if exogenous nationalist of regionalist factors were to prevent or even to delay comparative analysis of various TQM models. Ideally, debate should be allowed to continue freely for the purpose of defining a general TQM model (not a standard, as some people insist on) as the basis on which every company could build its own preliminary TQM model, which would then gradually be personalized to take account of the company's situation, goals, and experience. At the same time, award models should reflect the objectives and preferences of the organizers, with the greatest possible freedom to diverge from the general model and from the models used by other awards, according to the elements the award is intended to encourage and recognize.

This is the situation as regards TQM models, that is, the set of criteria, or categories, used to assess company quality (for example, the seven categories of the MBNQA or the 10 of the Deming Prize).

As far as the assessment method is concerned, that is, the way the assessment is performed,

- Two quite separate methods can be used: one begins with results, moves back through process, and, where possible, arrives at the quality system; the second proceeds in the opposite direction.

- The first method, which is typically diagnostic, must be used for improvement-oriented self-assessments.

- The second method should be used to verify whether the company's actual situation is aligned with its goals (management audits or external assessments in relation, for example, to a quality award TQM model, or checks to assess whether an application report reflects the real company situation).

The assessment method is another issue on which free discussion should be encouraged, particularly with the organizers of today's awards. Use of the award method (the second) for self-assessment should no longer be recommended. But the assessment method adopted by awards needs to be reviewed, too. Its principal aim should be to ascertain if and how the company being examined for the award performs its diagnostic self-assessments as part of its own continuous improvement process. Before self-assessment and quality awards are examined in detail, they are considered in the general context of company quality assessment.

Different Types of Quality Assessment

A quality assessment of a company or part of a company may be conducted for various reasons.

1. The assessment is commissioned by the customer, who needs to acquire confidence regarding the supplier's ability to meet his or her quality requirements. This is known as second-party assessment, because the party that commissions and owns the results of the assessment is a customer of the company concerned. The company is assessed in relation to a customer-specific or a sector-specific standard, or in relation to a general standard (for example, the ISO 9000 standards).

2. The company submits to an assessment by a third party, that is, by an independent agency which does not answer to either the customer or the supplier. The company may request the assessment in response to customer demand or on its own initiative, in anticipation of likely market requirements. Certification of conformity with international ISO 9000 standards comes under this category.

3. Top management wishes to assess the company's situation as regards quality. This may be because it intends to implement a total quality program and needs to know where to begin; or because a program has already begun but

an independent progress report is required, or, more simply, because management wishes to conduct a preliminary internal audit prior to a second- or third-party assessment. This is a first-party assessment and can be conducted by the company's internal resources or by an external agency. In this case, top management is the party that commissions and owns the results of the assessment. President's audits come under this category.

4. The assessment is an integral part of the corporate improvement process. This is self-assessment, a fundamental stage in the road toward total quality and the starting point for strategic improvement planning.

5. The company submits voluntarily to an independent third-party assessment in order to gain public recognition of excellence (national and international quality awards).

Assessment criteria vary greatly depending on the purposes of the actual assessment and the aims of the assessment owner. For this reason, the tendency to extend the use of criteria suitable for standards-conformity certification to total quality and national quality awards is extremely dangerous.

For the sake of simplicity, not all the categories listed previously will be discussed here. The most important categories for the present and near future are (2) and (5) – typical third-party assessments, also referred to as *external assessments* – and (4), self-assessment, also referred to as *internal assessment.* In the proactive context of total quality, self-assessment is the cornerstone of improvement, the vital element to which all other kinds of assessment must refer if they are to guarantee reliability.

The Company Model in Relation to the Quality Mission and the Assessment Reference Model

Assessments should be based on a model of the company and of the environment with which it interacts (suppliers and customers in the broad sense) – not a general model, but a model geared to the quality mission. In the age of total quality, the quality mission can be defined as "generating the maximum satisfaction of all users with the minimum use of resources; continuously improving this performance over time." The term *users* is employed here in the broad sense, so that the definition also covers the satisfaction of the company's employees and the beneficiaries of the company's results.

Figure 17.1 offers a synthetic representation of the model assumed here.[2] For simplicity, suppliers are included in the company's quality system/processes. Users are those shown under "user results" in Figure 17.2, in other words, everyone with legitimate expectations of the company (customers, employees, shareholders, society in general), but also everyone whom the company wishes to conquer (prospective customers). In this model, the significant element of the "users" block is represented by "expectations." The company responds to these expectations by activating "processes" which generate "products" (in the broad sense: everything the user sees as a response to his expectations, including attitudes). The point at which "products" meet "expectations" is the source of "customer satisfaction." In Figure 17.1, the term *results* indicates end results or user results, as measured by "customer satisfaction." Internal results, that is, the company's measurements of its processes/products before delivery to the user, are included under "processes."

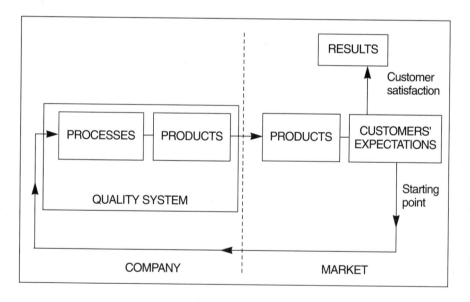

Figure 17.1.   The basic company model.

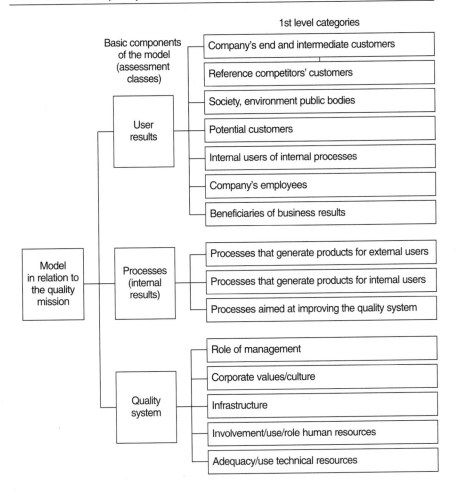

Figure 17.2. The company model as a tree chart.

Standards-Conformity Assessment and Assessment in Relation to the Total Quality Mission

Figure 17.2 presents the company model as a tree chart, down to the second-branch level. The model in Figure 17.3 reflects the concept that company quality assessment involves assessment of all three components in Figure 17.2 – results, processes, and quality system – and an appraisal of the consistency of the results of the three assessments.

Results are the basic component of the general assessment model in Figure 17.3. They provide irrefutable proof of the company's capabilities, of the progress made toward total quality, and of improvement trends. They also provide an accurate basis for verifying the other two components, which must be consistent with results. The logical assessment sequence moves from results to the processes that led to those results, and from there to the quality system beneath those processes.

All three components are therefore necessary for a correct assessment. Nevertheless, in certain instances, assessments are concerned only or mainly with the quality system (see Figure 17.4). This type of assessment, which has proved its historical worth and continues to play an important role today because it includes ISO 9000 certification, is covered by categories (1) and (2) in the previous section.

The simplified assessment model in Figure 17.4 is based on the following assumptions.

1. The specific purpose of the assessment (second- or third-party) is to give users confidence about the company's future results through a prior evaluation of the company's ability to deliver products that meet user requirements. By definition, this is a situation in which results are not – or may not – be present, so the assessment necessarily focuses on *how* the company ensures the quality of its results (this *how* is usually illustrated in a quality manual, which has to meet certain standard requirements).

   Assessments of this type are frequently an entry requirement for new suppliers. Once the supplier has been formally approved, a full capability assessment is performed, through the use of vendor ratings. In this case, too, results are necessary for a complete assessment.

2. The assessment is concerned with only one area – even if multifunctional – of the company quality system, the area that generates products. This is the area with which the ISO 9001/2/3 standards are concerned, and it is the most tangible and easily verified part of the company quality system. In fact, some of the processes involved (manufacturing processes) can be kept in a state of statistical control and their results are statistically predictable. In this case, it is considered legitimate practice to infer conclusions regarding product quality from the assessment of the quality system (in particular, from process capabilities).

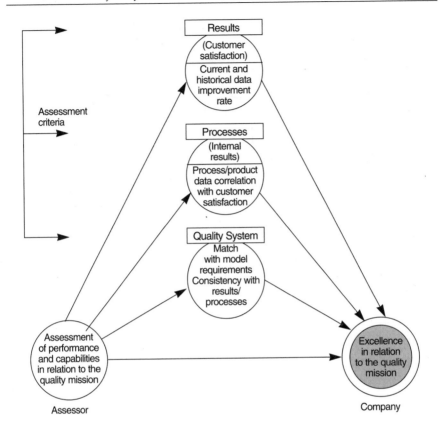

Figure 17.3.   Company quality assessment model – general model.

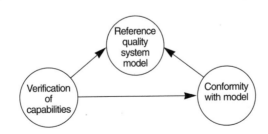

Figure 17.4.   Company quality assessment model – simplified model.

Great care must be taken, however, to guard against the tendency to extend the logic on which the simplified assessment model of Figure 17.4 is based to contexts in which these considerations no longer apply. Extending this logic means not only applying the model in Figure 17.4, but also assigning excessive weight to the quality system in Figure 17.3 (the system's capabilities), to the detriment of results. In some award models, the quality system has a much greater weighting than results. The suspicion inevitably arises that long familiarity with second- and third-party assessments for standards conformity lies at the root of the priority given to the quality system, or capabilities, over results.

When the purpose is self-assessment or assessment for an award [categories (3), (4), or (5) of the previous section] rather than a prior assessment of capabilities [categories (1) and (2)], external results (from the user perspective) and internal results (processes/products) must evidently be available (consideration 1 is therefore not valid). Moreover, as the quality system becomes more complex and the number of intangibles rise, consideration 2 becomes increasingly arbitrary. Although many people refer optimistically (or rashly) to six sigma (in other words, to situations in which processes are assumed to be in a state of statistical control) for processes in which human, social, communication, organizational, and political variables are significant factors, it is best to avoid all forms of determinism, even of a probabilistic type, where total quality is concerned. In the present stage of development of total quality models, a positive match between the actual quality system and the theoretical model is, at best, a necessary though certainly not sufficient condition to ensure reliable results.

Inspection-Oriented Assessments and Improvement-Oriented Assessments

To sum up so far, the closer the company moves toward total quality the lower the objectivity of the quality system model and the smaller the possibility of extrapolating results from the degree of conformity with the model. This means assessments must have access to real, appropriately weighted results. A good balance must be established between the three components (quality system, internal results, and external results).

A second requirement is that the three components must be consistent with one another. A lack of consistency indicates incorrect assessment/measurement of one or more components. The three components are in fact intrinsically linked with one another in a cause-effect relationship (apart from time scale discrepancies, for which allowance must be made). The fact that these correlations are often not obvious – or, given their complexity, not sufficiently obvious for qualitative deductions to be made – does not mean they do not exist. On the contrary, an analysis of the cause-effect chains that emerge from assessments will bring these correlations to light and enable the company to improve both its process indicators (for better alignment of internal results with user results) and its quality system model.

Verifying the consistency of the three components is a fundamental part of the assessment process. Verification can begin with the quality system (categories and items of the award models, or the five categories of the third branch of the model in Figure 17.2) and end with results. Alternatively, consistency can be verified in the opposite direction, beginning with user results and then moving backwards through the generator processes to the quality system. In this way, the cause-effect chains are followed through, beginning with effects (see Figure 17.5).

The first type of assessment begins with the quality system, more precisely with a reasonably accurate model of the quality system (possibly just a working hypothesis). Since the MBNQA is an important example of an assessment approach which begins with a quality system model (a conceptual model, not a prescriptive model for implementation), it is used for reference here (see Figure 17.5a). For each item in the model, the degree to which the actual situation conforms with the requirements of the model is checked to assess the company's specific approach; next, the degree to which this approach is applied within the company is verified (deployment); finally, results are assessed. But results are not necessarily user results. Results exist at various levels of the cause-effect chains and the assessor will often be forced to end his or her appraisal at an intermediate, internal-results level, and be unable to reach the user results at the ends of the process chains.

The second assessment approach begins with results and works back along the chains to the quality system (see Figure 17.5b). In other words, it begins with data that, by definition, are real – user results – rather than with models of reality, and its purpose is diagnostic: to identify the causes of deviation from expected results and therefore pinpoint the weaknesses and strengths of the company's processes and quality system in order to permit adjustment and improvement.

As noted earlier, results are the pivot of the model assumed here (based on a customer-centric rather than company-centric perspective). This produces a strategic but at the same time concrete approach to improvement (results-driven improvement). The two assessment methods illustrated in Figure 17.5, the left-to-right

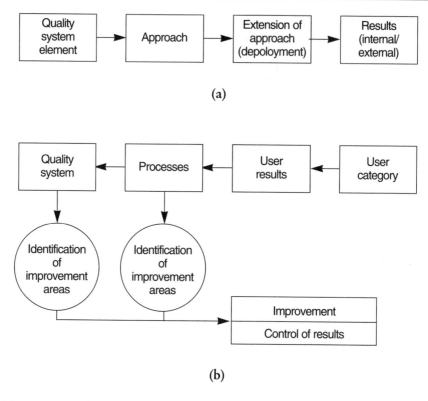

Figure 17.5.   The two appraches to assessment: (a) from the quality system to
results (control-oriented assessment); (b) from results to processes to
the quality system (improvement-oriented assessment).

approach and the right-to-left approach respectively, are not equally valid alterna-
tives in the case of improvement-oriented assessments; in this case, only one option
is open, the right-to-left method.

Only the typically diagnostic right-to-left method (Figure 17.5b) highlights
inconsistencies in process indicators (incorrect alignment between internal and
external results), shortcomings in process capabilities, and flaws in the quality system (missing or insufficiently weighted items). Above all, it enables the company

to build a customized quality system model, to gradually move away from the usually general-purpose model assumed beforehand, such as a quality award model, toward a specific corporate model. In other words, the criterion of continuous improvement based on reality, on facts (and results are facts, the quality system is just a model of the causes), is applied to the quality system, too.

For these reasons, the assessment illustrated in Figure 17.5b is an improvement-oriented assessment. Inspection-oriented assessments are perfectly adequate for checking that everything is going according to plan, in particular that the quality system and process improvement planned after an earlier diagnostic self-assessment have been implemented. Therefore, the methods that are suitable for external award assessments are not equally suitable for self-assessments. This is a conclusion of some significance, in view of the extremely widespread use of the award models for improvement-oriented self-assessment. The risk of a model-driven view of improvement developing rather than a results-driven view is all too real.

Assessment Reliability

It is also useful to examine the question of assessment reliability. The first point to be made is that reliability depends closely on the range and nature of the quality characteristics being assessed. It decreases as the range widens and quality becomes increasingly total.

However, the presence of user results and process/product measurements significantly raises assessment reliability. The quality system is always the most critical component. Even at the level of product generation, the most tangible and easily controlled subsystem, the quality system is a critical factor whose weight increases exponentially as the system broadens to embrace the entire company and a growing number of intangibles.

A second factor affecting reliability is who controls the assessment – who commissions and owns it. A self-assessment commissioned by the company for its own purposes to monitor its situation and plan improvement is by far the most reliable assessment. The company's aim is to understand, it has nothing to gain by creating ambiguities or distortions.

This is the aim of the company, but not necessarily of those delegated to perform the assessment or of those asked to express their opinions. The company must therefore adopt a methodology that will guarantee the greatest possible impartiality.

As a general rule, external parties should be appointed to conduct the most delicate phases of the assessment – such as the interviews – and to evaluate the main intangibles. This will create an image of impartiality and confidentiality and ensure that the assessment is conducted by people with the necessary skills and experience.

The reliability of assessments by second parties (customers) and third parties (independent certification organizations or independent award-assigning organizations) is low. The company tends to assume a defensive attitude in second-party assessments. If the external assessment factor is combined with the increase in complexity factor, reliability drops sharply. This is why the utility of extending to total quality those methods that already involve an element of risk when applied to less complex situations (certification) is questionable.

How, for example, can intangibles (such as leadership, customer culture, the team spirit, and respect for people), which are such important factors in determining corporate excellence, be assessed on the basis of a report submitted by the company's management (the application reports used by the awards) and a three- or four-day site visit by a group of examiners?

Reliability also depends on a third factor: the people who make judgments. Here, the situation varies according to which of the model's three components is being assessed: results, processes, or the quality system.

No difficulties arise as far as results are concerned: Opinions are expressed by the customer (in the broad sense). Customers have expectations and are therefore qualified to judge the degree to which products (in the broad sense) meet their expectations. This is why results are considered the most reliable data and why it is suggested that a very high, predominant weight should be given to results if an award is to be credible.

For processes/products, assessment reliability is potentially high because it deals with measurements. In practice, measurements are reliable only if the process is in a state of control and aligned with user results. Since the voice of the processes is indispensable for a serious assessment, careful attention must be paid to the state of process management within the company.

Assessments of the quality system are conventionally made by management, in reports frequently based on audits. But audits are potentially objective only when they deal with tangible characteristics. Where intangibles are concerned, the assessment is often opinion-based, and the opinion of management tends to be optimistic.

Management, in particular top management, is the supplier and the quality system and as such is inclined to have a favorable opinion. Management also tends to believe that the situation is as it should be and not as it is, particularly in the case of the main intangibles, with which management is often directly concerned (leadership, the atmosphere in the company, decision-making).

Assessments of the quality system are much more reliable if the system users (company personnel at all levels, from blue-collar workers to management) are asked to express a judgment. It must be made clear that the assessment is not an internal user satisfaction survey, but a fitness-for-use survey in which employees are the users of the quality system. This approach is consistent with the quality philosophy, which evaluates every product in terms of its fitness-for-use.

Assessment by the user also goes a long way to solving the problem of assessing intangibles. As a group, the people who work in the company have an intimate understanding of its intangible characteristics, even if the mathematical skills of external specialists are usually needed to collect this information in a suitable form.

To conclude, assessment reliability requires judgments from a variety of sources, expressed in formal management reports, audits (for tangible characteristics), and employee questionnaires/interviews (for less tangible characteristics).

The Central Role of Self-Assessment

Self-assessment plays a central role in the improvement-planning process. It was defined earlier as an assessment activity performed at the wish not of a customer or any other external party, but of the company itself led by its top management. A word of warning, however: Self-assessment does not signify that the assessment reflects the opinions of management, expressed directly or through the company staff. The task of management is to get others to speak – users in particular – and to listen to them. Nor does it mean that external parties are excluded from the process. On the contrary, external parties are usually in the best position to monitor the voice of the users without influencing them and to perform specialist assessments, but they are working for the company, operating as experts in a company process.

The model described in the previous sections indicates the input for the self-assessment process. In Figure 17.6, input 1 represents the results of customer satisfaction surveys of all user classes, input 2 represents the results of current measurements and periodic audits of the company's main processes, and input 3 represents the results of audits and fitness-for-use surveys on the quality system.

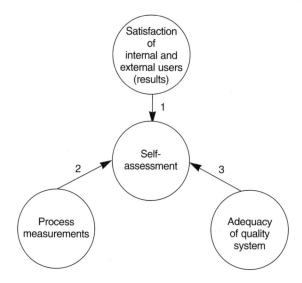

Figure 17.6.   The self-assessment process.

Self-assessment should also play a central role as a compulsory basis of reference for external assessments, especially assessments of excellence in the field of total quality. This is chiefly for practical reasons. If it is true that results and processes are always vital to ensure reliability, above all in total quality assessments, then no external assessor could afford to conduct an independent survey of customer satisfaction, nor could it check the reliability of process measurements, because that would require precise audits of the state of control, or verify the alignment with user results. The external party needs information summaries, comparisons, and assessments of results and processes, which can only be obtained through a systematic, complete, and detailed self-assessment procedure conducted for the company's own purposes.

Moving from results and processes to the quality system, the low reliability factor of external assessments becomes predominant. Even in simple assessments concerned solely with the product quality assurance subsystem, defensive attitudes or the exam syndrome can impede understanding. When the quality system is extended to the entire company and intangible characteristics predominate, it seems unwise to imagine that external assessors can understand and assess the state of the system on the basis of a report by the company's management or a site visit. Similarly, it would be unwise to count on the frankness and objectivity of people whose sole aim can only be to pass the exam.

This is the Achilles' heel of some awards, which consider the application report a self-assessment simply because it was drawn up by the company itself, naturally on the basis of all available data, when in fact it is an ad hoc report, prepared with the intention of competing for the award.

Self-assessment as intended here is entirely different. It is a much broader, more detailed process than that set in motion by an ad hoc report (which has to follow the guidelines laid down by the external organizations and not exceed a certain number of pages). Above all, it is an integral part of the company's total quality process. All external assessments should be based on the self-assessments performed by the company (improvement-oriented self-assessments, which begin with results). In the case of awards, the availability of self-assessments (referring to at least the previous two years) should be a requirement for companies applying for an external assessment.

Quality Awards: The Positioning Problem

The positioning concept is useful to shed greater light on the specific characteristics of an award. It helps to clarify the intentions of the award and its particular features, which usually are not explicitly stated and do not emerge clearly from the award documentation.

The first positioning concerns the comprehensiveness of the quality system, that is, the quality system's proximity to the ideal limit of total quality. The traditional product assurance quality system can be positioned somewhere toward the lower end of the scale. The upper end is an ideal quality system covering the entire organization and everyone working in it; in other words, the system that responds in full to the total quality mission. In Figure 17.7, this positioning is represented by the horizontal axis.

The Deming Prize, the MBNQA, and the European Quality Award can all be placed in the same area as regards the comprehensiveness of their quality systems.

This positioning is therefore not a differentiating factor. The three awards, in the current phase of quality development, are well positioned as regards total quality. If they were placed further toward the right, assessment would be even more complex and less reliable.

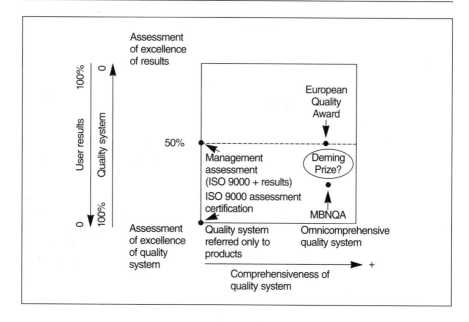

Figure 17.7. The positioning of assessments and awards.

The true differentiating factor of the three awards is provided by the second positioning, between quality system and results (vertical axis in Figure 17.7). This positioning is best understood by considering the extreme case of an award focused solely on company excellence in achieving customer satisfaction: This award would be at the top end of the scale. In theory at least, another award might be geared solely to the excellence of the company's quality system and would be placed at the bottom end of the scale (ISO 9000 assessments in Figure 17.7).

The key role of results in any assessment has been stressed repeatedly, but a conceptual distinction needs to be drawn between self-assessment and external assessments. In the case of self-assessment, the results are vital because they represent the starting point of the assessment and the yardstick for verifying the other two components: processes and the quality system (results-driven assessments oriented to improvement). In the next stage of self-assessment, however, the improvement planning and implementation stage in which the company aims to create the conditions for better future results, the roles are reversed (that is, the focus shifts from results to the quality system and processes).

In the case of external assessments, results must always be predominant, not only to ensure assessment reliability, but because their presence and use demonstrate that the company is moving correctly toward total quality (regularly performing diagnostic assessments). In the total quality field, external assessment of the quality system is such an uncertain business that it is always best, even for awards for the excellence of the company approach, to wait until excellence is reflected in results, rather than risk giving an award to an apparently perfect system, which then fails to produce the expected results.

As far as vertical positioning is concerned, the European Quality Award is located in the middle of the scale, with a 50 percent weighting to the voice of the users (results) and a 50 percent weighting to the voice of the company (quality system + processes). Despite the apparently high weight given by the MBNQA to the customer satisfaction category (30 percent), only 15 percent is actually given to results (7.5 percent each for the company's own results and results as compared with competitors). The positioning of the Deming Prize is not clear, since the weights given to the assessment categories and items are not known.

The second positioning offers significant clues to the objectives of the awards. The primary aim of the European Quality Award clearly is to give ample scope to users in assessing company excellence and to give credit to a TQM system that has already proved its effectiveness. The chief objective of the MBNQA seems to be recognition of the excellence of the quality system: indication of national TQM models which can be imitated (with the risk, explained earlier, of some inconsistency with results). Since the vertical positioning of the Deming Prize is not clear, it is difficult to form opinions on this award's primary objective. However, the origins of the prize and the information available suggest that, like the MBNQA, the chief purpose of the Deming Prize is recognition of the excellence of the TQM system; there is reason to believe, nevertheless, that in practice, if not in theory, significant weight is given to results.

It should be evident from this brief analysis that a clear statement on positioning would facilitate a better understanding of the specific characteristics of quality awards.

Making the Appropriate Distinctions Between Awards and Self-Assessment

The following conclusions can be drawn.

1. The method. The generalized use for self-assessments of the left-to-right method (approach-deployment-results) adopted for award assessments should no longer be recommended. The left-to-right method is suitable for the preliminary inspections performed by companies wishing to compete for awards before undergoing external assessment, and for inspections to ascertain the match between the specific company model described in the application report and the award's TQM model and the match between the company's real situation and the description of this situation in the application report. It can also be used for the checks the company performs on its own behalf to assess the conformity between its real situation and its models and improvement plans.

   But when improvement is the company goal, self-assessment must adopt the diagnostic right-to-left approach. This is the self-assessment method that should be promoted, for the good of companies and for progress in quality; this is the self-assessment method which awards should take as a premise and on which award assessments should focus for an accurate insight into the company's results, processes, and quality systems.

2. The model. Research into total quality models should continue. Indeed, the development of a general TQM model to be used as the framework and foundation for specific models should be encouraged. In the next few years, a general TQM model may become established; or a variety of models may be maintained. Whichever the case, the model should not be regarded as a standard. This would impede the continuous development of quality and the tendency to use the model as a basic menu to be personalized according to specific requirements.

The question of a general model is discussed elsewhere.[3] The general model should be similar to that illustrated in the tree chart in Figure 17.2, with the quality system expanded to include the elements shown in Figure 17.8. The term *quality system* indicates all those characteristics of the company that play a role in the fulfillment of the quality mission as defined earlier.

The tree obtained contains all the elements needed to assess the company according to the assessment model in Figure 17.3 (and if the tree is complete, these elements will be sufficient).

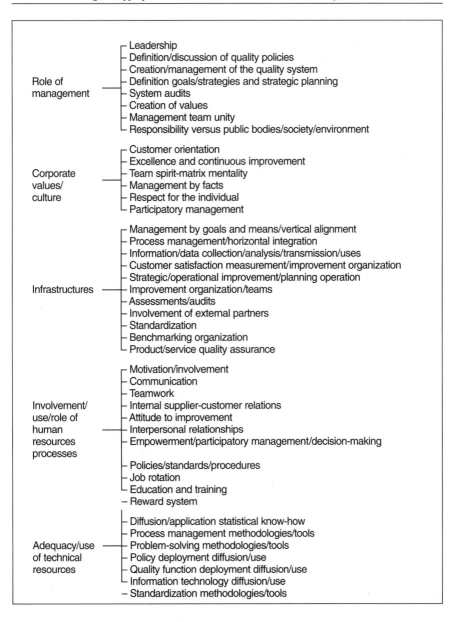

Figure 17.8. Example of a second-level deployment of the quality system.

It is stressed that the general model obtained by combining Figure 17.2 and Figure 17.8 is simply an example, a useful working hypothesis for the purposes of this discussion and a basis for a debate on the general TQM model. Here, it is assumed as a working hypothesis and starting point both for self-assessments and for awards.

Awards can be considered first. An organization that intends to set up a new award can take the general model as its basis of reference, as a menu of all the possible items. According to the main objectives of the award, it will define its positioning by allocating appropriate weights to the main branches: external results, internal results, and quality system (see Figure 17.2). Weights will then be allocated to the second-level branches in Figure 17.2 and to the third-level branches pertaining to the quality system in Figure 17.8, according to the award's specific objectives. For example, the overall weight of the first-level branch results can be divided among the different users (second-level branches) to achieve the finest distinctions, in line with the requirements of the market, the environment, and the workforce. The allocation of weights to the subdivisions of the quality system permits even greater differentiation. This type of differentiation is useful at the level of national/international awards, but it is essential when the award is concerned with a more specific area, for example, a particular business sector. Awards set up by associations of companies in the same merchandise area or market sector come under this category.

Now we can consider self-assessment. If there are no external constraints, the company can personalize the general model exclusively in relation to its own requirements: its market, the environment, the situation of the company, its culture, its values, and so on. Personalization means eliminating irrelevant items and attributing weights to those that remain. Even if the company decides to work with an external constraint – for example, its starting point is the national/regional/sectional award model – the general model will provide a basis of reference to help the company identify the different assessment categories and personalize the preliminary model to meet its particular needs.

It cannot be emphasized too strongly that too much is at stake to permit self-assessment to be subordinated to concerns outside the company's sphere of interest. Total quality is an abstract concept until it is molded by the specific reality of the individual company. The advantage of the model is that it provides a framework. Specifically, the general model is a starting point from which sectional models for homogeneous groups of companies can be developed. The final phase, the most important and delicate, can only be taken by the individual company, in its pursuit of excellence: adaptation to the particular situation of the company, which is a unique reality. In fact, it would be more appropriate to refer to this phase as a process that never ends.

In the next few years, it is to be hoped that

- Research into a general TQM model(s) will continue, unconditioned by quality awards
- Companies in the same merchandise areas/market sectors will cooperate in adjusting the general model to the particular requirements of the individual sector, to create sectional models which each company can use as its starting point
- Individual companies will regard self-assessment and the continuous monitoring, checking, updating, and personalizing the model as the key to total quality
- Where awards are concerned, the use of a general reference model will make it easier to identify the specific objectives of an award and to position it clearly

Once award organizations abandon the idea that their award should provide the best TQM model possible, it will become easy to discard religious-dogmatic attitudes to awards and to pursue freely the objectives dictated by specific geographic, social, industrial, and market situations.

Notes

1. Tito Conti, "Company Quality Assessments," *The TQM Magazine,* June 1991 and August 1991.
2. For a full discussion of the model, see: Tito Conti, *Building Total Quality* (London: Chapman & Hall, 1993), chapter 2.
3. Ibid.

Section II:

Other Papers Dealing with Quality Awards

Chapter 18:

The Australian Quality Awards

THE AUSTRALIAN QUALITY AWARDS FOUNDATION

In 1988, as part of its Australia for Quality national awareness campaign to promote the total approach to quality, Enterprise Australia introduced the Australian Quality Awards to

- Recognize organizations that demonstrate outstanding achievements in the improvement of the quality of their products, services, and management

- Encourage others to follow their example

- Raise the level of awareness in the community, and in business in particular, of the importance of quality to competitiveness and the standard of living

From the outset the awards have had the personal endorsement of the prime minister and senior ministers. They have also enjoyed strong support from sponsor organizations that have provided the significant financial backing that has made the awards possible and from the many companies who have released senior managers to participate as evaluators in the judging process each year.

The awards have grown in stature and significance over the years since their inception. They are now acknowledged as a reward for excellence, and the evaluation criteria employed are used by a growing number of organizations to help monitor and assess their progress. They have contributed to the growing awareness of the importance of quality in Australia and to the improved quality observed in many organizations.

However, while many Australian companies have shown significant improvement in recent years, international competitors have commonly matched or exceeded those gains. The Australian Quality Awards Foundation was formed in 1991 in recognition of this and to provide for the expansion and funding of the awards.

The Australian Quality Awards and the U.S. MBNQA were introduced at the same time and, from the outset, have independently adopted very similar criteria to assess performance.

The presentation and interpretation of the criteria were modified in 1991 to facilitate international comparison and assist companies that wish to relate to other awards programs. The 1992 criteria continued this approach and, in 1993, as in previous years, Australia has slightly modified the emphasis in certain of the criteria, and the weighting of those criteria, in the light of experience and changing circumstances. Awards are given in various sections to organizations that achieve prescribed levels of performance.

There is no restriction on the number of awards that may be won in any section. The judges may decide to make no awards if they consider the standard of the entries does not warrant it.

Benefits of Participation

Winning an Australian Quality Award carries with it national recognition of an outstanding achievement in the pursuit of quality excellence. The awards have been presented, in previous years, by the prime minister and senior federal ministers, reflecting their acknowledgment of the award's national significance.

In 1993 the awards presentation will take place at the World Congress Centre in Melbourne on November 23, 1993. It is planned that the proceedings will again be linked to similar gatherings in each of the capital cities. It is anticipated that the prime minister will present the awards and that the state premiers will attend.

The publicity accorded to the awards and the winners increases each year. It is anticipated that there will be extensive television, radio, and press coverage of the awards, including special press supplements and coverage.

A summary of each winner's entry will be included in a book of case studies to be published in 1994. Finalists may be invited to make presentations to the 1994 Australian Quality Awards Tours and to other seminars and public forums across the country.

Award winners will receive an award statuette in perpetuity. They are entitled to use the awards symbol and to refer to the winning of the award in their own advertising.

There are major benefits for all entrants from participating in the awards process, including

- The impact of a clear public commitment to quality excellence
- A heightened awareness of the importance of quality and a fuller understanding of what it entails
- A focal point for quality activity
- A structured and systematic self-assessment of performance (required by the application)
- An external evaluation and reporting process that provides applicants with an objective appraisal of their performance

- An encouragement to greater teamwork, cooperation, and pride in performance, throughout the organization

Participation in the awards is also a prerequisite for entry to the Australian Quality Prize program.

Previous Awards Winners

The Awards for Product Quality Improvement, Service Quality Improvement, and Individual Quality Improvement Projects, which were offered previously to encourage wider participation in the early stages, were discontinued in 1991.

Awards for Outstanding Service Quality Improvement were made to Control Data Australia Pty. in 1988 and Rank Xerox in 1990. No award was given in 1989.

Awards for Outstanding Product Quality Improvement were won by W.A. Deutsher Pty. Metal Products Division in 1988 and Kodak (Australasia) Pty. in 1989. No award was given in 1990.

Awards for an Outstanding Individual Quality Improvement Project were made to Darling Downs Cooperative Bacon Association in 1988, Girvan NSW (Citadel Towers Project) in 1989, and Alcoa of Australia, W. A. Bauxite Mining Operations in 1990.

Past winners of the Australian Quality Award for Outstanding Achievement in TQM were Kodak (Australasia) Pty. in 1988; Ford Motor Company of Australia, Sydney Assembly Plant, in 1989; and BHP Steel Slab & Plate Products Division in 1990.

Since 1991 the Australian Quality Award has been offered only for organization-wide quality improvement and in three sections. Winners in 1991 were

- In the large organization section: Baxter Healthcare Pty., The South East Queensland Electricity Board; Toyota Motor Corporation Australia, Tubemakers of Australia.

- In the subsidiaries and divisions section: BHP Steel High Carbon Wire Products Division

- In the small enterprise section no award was made

Winners in 1992 were

- In the large organization section: The National Roads and Motorists Association (NRMA); Avis Australia

- In the subsidiaries and divisions section: the Ford Motor Company of Australia, Plastics Plant

- In the small enterprise section no award was made

The inaugural Australian Quality Prize was awarded in 1992 to Kodak (Australasia) Pty.

The 1993 Awards Process

Eligibility

The awards are open to the following organizations.

- Companies incorporated, and/or physically operating, in Australia

- Commonwealth and state government departments and instrumentalities, and local government bodies

- Such other organizations as the awards committee may from time to time deem eligible to apply

Participants may enter one of the following sections.

1. **Large Organizations** (fully autonomous)
   To be eligible for this section the organization must directly employ more than 250 full-time people and exercise the full range of management responsibilities required, including finance, administration, legal, manufacturing, personnel, distribution, research, sales, and marketing, as appropriate.

2. **Subsidiaries and Divisions** (not fully autonomous)
   A subsidiary or operating division is defined as part of an organization in which at least some of the decisions are made elsewhere, such as at the corporate level. However, it should be a largely autonomous business unit with its own senior management group responsible for a wide range of management activities. Clearly the scope of such responsibilities may be less than those for the parent organization. Government departments and utilities should interpret these requirements in a way appropriate to their function, seeking specific guidance from the foundation if necessary.

   A subsidiary or division must be able to demonstrate activity across the full spectrum of management by addressing all the categories and items of the evaluation criteria. In some cases this may involve discussing aspects of management that are not under the full or direct control of the subsidiary or division. Evaluation of the application, including the site visit stage, will include all of these aspects.

Such organizations are eligible to enter this section if they directly employ 150 or more full-time persons.

Organizations seeking entry in the subsidiaries and divisions section are advised to seek guidance on eligibility from the foundation at an early stage. The name and structure of the subsidiary or division and its relationship to its parent organization must be clearly stated in the application form. This statement must also include a description of how important management functions not under the full control of the subsidiary or division are supplied by its parent or some other organization.

As a rule a subsidiary or a division is not eligible to enter this section if its parent organization and/or other subsidiaries or divisions of the parent organization directly receive or deal with 50 percent or more of its products or services. However, exemption from this condition may be given in certain circumstances.

3. **Medium Enterprises** (fully autonomous)
To be eligible for this section the organization must directly employ more than 100 but less than 250 full-time people and exercise the full range of management responsibilities required, including finance, administration, legal, manufacturing, personnel, distribution, research, sales, and marketing, as appropriate.

This section was introduced in 1993 to enable the foundation to modify the eligibility criteria for small enterprises and thereby to better cater to their needs.

4. **Small Enterprises** (fully autonomous)
To be eligible to enter this section the enterprise must directly employ more than 20 and fewer than 100 full-time people and exercise the full range of management responsibilities appropriate to its operations.

Eligibility requirements for this section have been modified in response to input from the small business community. It is recognized that application for the awards as they are currently structured is proportionately more difficult for a small enterprise, particularly for organizations with 50 or fewer employees. It is planned to fully review the eligibility requirements and application process for the small enterprise section in 1993 with the assistance of interested parties.

Multiple Applications

A maximum of two subsidiaries or divisions of the same parent organization may apply for awards in the same year.

Subsequent Applications

An organization that has won an Australian Quality Award may not again enter for the award within three years of winning.

The winner of an award in the large organization or medium enterprise sections may not enter one of its subsidiaries or divisions in the subsidiaries and divisions section within three years of winning an award.

Winners of an Australian Quality Award are encouraged to apply for the Australian Quality Prize.

Eligibility Determination

It is a requirement that all applicants seek clarification of their eligibility to apply before submitting a full application. Potential applicants will be able to attend half-day seminars designed to explain the application process in more detail. A small fee will be charged.

Evaluation and Site Visit Fees

Large Organizations

Organizations with less than 3000 employees will be charged an evaluation fee of $2000, plus an additional fee of $1000 should the applicant be selected for a site visit. Organizations with more than 3000 employees will be charged an evaluation fee of $3500, plus an additional fee of $1500 should the applicant be selected for a site visit.

Subsidiaries or Divisions

Organizations with less than 3000 employees will be charged an evaluation fee of $1500, plus an additional fee of $1000 should the applicant be selected for a site visit. Organizations with more than 3000 employees will be charged an evaluation fee of $2500, plus an additional fee of $1500 should the applicant be selected for a site visit.

Medium Enterprises

Organizations will be charged an evaluation fee of $1000, plus an additional fee of $1000 should the applicant be selected for a site visit.

Small Enterprises

Organizations will be charged an evaluation fee of $500, plus an additional fee of $1000 should the applicant be selected for a site visit.

Payment of Fees

- Advisory Seminars: The cost and location of the seminars will be advised on notification of eligibility to apply, or on application to the foundation

- Evaluation Fee: Payable on forwarding the detailed submission

- Site Visit Fee: Payable on notification that a site visit is to be conducted

Use of the Guidelines

These guidelines are designed for use by an organization making an award application. They are also designed to be used as an internal improvement audit tool by Australian organizations. The awards process provides a framework for organizations to develop international best practice. In recognition of this, the evaluation criteria are being used by an increasing number of organizations to assess their own progress in the implementation of quality. The foundation encourages their use in this way, and facilitates such use by providing a series of public awareness seminars and specific training courses for internal evaluators. However, as the full awards process includes a valuable external assessment, organizations contemplating the use of the criteria for internal audit are strongly advised to consider it as a first step toward an application for the awards.

Photocopying of the guidelines for specific use in support of an awards application or internal audit is permitted; however, additional copies are available for a small fee.

The Australian Quality Awards Foundation reserves all copyright relating to the guidelines, and its use for fee-based activity is not permitted without the specific agreement of the foundation.

For further information about seminars and internal evaluator training programs please contact

Australian Quality Awards Foundation Ltd
P.O. Box 298
St Leonards, NSW 2065
Phone: (02) 439 8200
Fax: (02) 906 3847

Additional Copies of the Guidelines

The initial copy of the application guidelines is available free of charge from the foundation. Additional copies are available for a fee which includes postage and handling charges.

- One to five copies – $12 per copy
- Six to 10 copies – $10 per copy
- 11 or more copies – $8 per copy

Please send your order, with payment, to the address listed previously.

Subsequent Publicity

Winners of the 1993 awards will be entitled to refer to their success in media advertising and promotional material or on stationery. Any such reference must be stated in the following form.

- **For large organizations:** Australian Quality Awards Winner 1993
- **For subsidiaries and divisions:** Australian Quality Awards 1993 Winner Subsidiaries and Divisions
- **For medium enterprises:** Australian Quality Awards 1993 Winner Medium Enterprises
- **For small enterprises:** Australian Quality Awards 1993 Winner Small Enterprises

The Australian Quality Awards are for organizationwide quality improvement. The logos and associated text are to be used only for general advertising and not as a specific product or service endorsement.

A winner in the subsidiaries and divisions section may use the logo only in relation to that subsidiary or division. The logo may not be used in such a way to imply that the parent organization, or the organization of which the winner is a part, has won an award.

The Australian Quality Awards winner logo must stipulate the year of the award on all publicity material. Award winners have the choice of identifying their status in print with the logo in 'PMS Gold 873' and black typeface, or black logo and black typeface on a white background.

No other reference will be permitted except with the written consent of the Australian Quality Awards Foundation.

The concept of quality as used throughout these guidelines means managing the total organization using quality principles. Quality is best defined by means of a number of key concepts rather than by a simple definition. These key concepts include, but are not restricted to

- Leadership to create and deploy clear values to the organization

- A community and environmental responsibility appropriate to the organization's activities

- A planned and structured approach to setting and achieving goals and objectives

- An understanding of variation and management by appropriate facts and data

- Fully involving and developing the organization's people

- The customer playing the central role in the definition of product and service quality

- The organization and its suppliers working in partnership

- Quality derived from well-planned and managed processes

- Standardization being part of process management

- Continuous improvement being part of the management of all processes

- Innovation being recognized as an essential adjunct to continuous improvement

- Management emphasis on prevention and improvement rather than reaction

The Australian Quality Awards evaluation criteria, summarized in Figure 18.1, reflect and expand these key concepts and are designed to allow a wide range of organizations, large and small, public and private, to concisely describe their own approach to quality. The emphasis of evaluation is on how well the organization mobilizes all of its resources and links and directs all of its activities toward achieving its goals. Applications must be made on the official application form. The form may be photocopied. The application form seeks basic information on the applicant; its organizational structure; its range of products, services, and trading activities; and the location of its offices and plants (see Figure 18.2).

The application form must be forwarded before the detailed submission so that the eligibility of the organization can be determined.

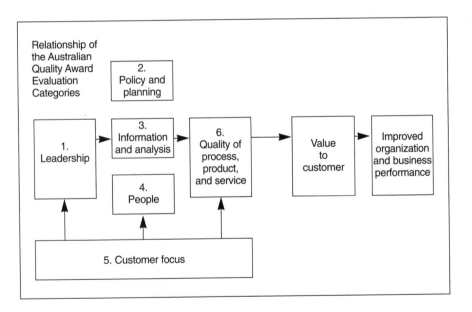

Figure 18.1. The Australian Quality Awards evaluation criteria.

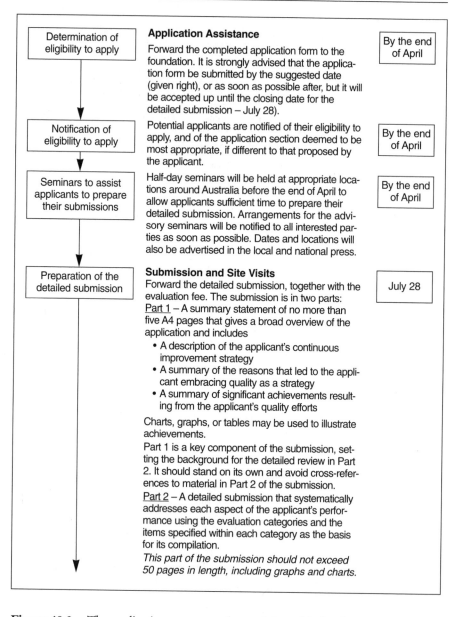

Determination of eligibility to apply

Application Assistance

Forward the completed application form to the foundation. It is strongly advised that the application form be submitted by the suggested date (given right), or as soon as possible after, but it will be accepted up until the closing date for the detailed submission – July 28).

By the end of April

Notification of eligibility to apply

Potential applicants are notified of their eligibility to apply, and of the application section deemed to be most appropriate, if different to that proposed by the applicant.

By the end of April

Seminars to assist applicants to prepare their submissions

Half-day seminars will be held at appropriate locations around Australia before the end of April to allow applicants sufficient time to prepare their detailed submission. Arrangements for the advisory seminars will be notified to all interested parties as soon as possible. Dates and locations will also be advertised in the local and national press.

By the end of April

Preparation of the detailed submission

Submission and Site Visits

Forward the detailed submission, together with the evaluation fee. The submission is in two parts:

Part 1 – A summary statement of no more than five A4 pages that gives a broad overview of the application and includes

- A description of the applicant's continuous improvement strategy
- A summary of the reasons that led to the applicant embracing quality as a strategy
- A summary of significant achievements resulting from the applicant's quality efforts

Charts, graphs, or tables may be used to illustrate achievements.

Part 1 is a key component of the submission, setting the background for the detailed review in Part 2. It should stand on its own and avoid cross-references to material in Part 2 of the submission.

Part 2 – A detailed submission that systematically addresses each aspect of the applicant's performance using the evaluation categories and the items specified within each category as the basis for its compilation.

This part of the submission should not exceed 50 pages in length, including graphs and charts.

July 28

Figure 18.2.   The application process and award timetable for the Australian Quality Award.

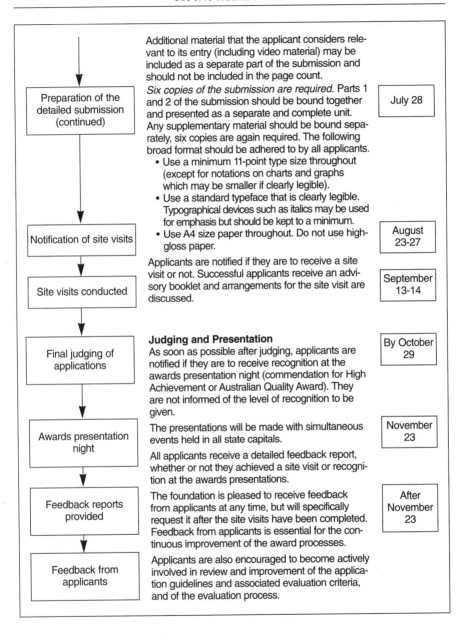

The flowchart contains the following boxes and text:

Preparation of the detailed submission (continued) — July 28

Additional material that the applicant considers relevant to its entry (including video material) may be included as a separate part of the submission and should not be included in the page count.

Six copies of the submission are required. Parts 1 and 2 of the submission should be bound together and presented as a separate and complete unit. Any supplementary material should be bound separately, six copies are again required. The following broad format should be adhered to by all applicants.

- Use a minimum 11-point type size throughout (except for notations on charts and graphs which may be smaller if clearly legible).
- Use a standard typeface that is clearly legible. Typographical devices such as italics may be used for emphasis but should be kept to a minimum.
- Use A4 size paper throughout. Do not use high-gloss paper.

Notification of site visits — August 23-27

Applicants are notified if they are to receive a site visit or not. Successful applicants receive an advisory booklet and arrangements for the site visit are discussed.

Site visits conducted — September 13-14

Judging and Presentation

Final judging of applications — By October 29

As soon as possible after judging, applicants are notified if they are to receive recognition at the awards presentation night (commendation for High Achievement or Australian Quality Award). They are not informed of the level of recognition to be given.

Awards presentation night — November 23

The presentations will be made with simultaneous events held in all state capitals.

All applicants receive a detailed feedback report, whether or not they achieved a site visit or recognition at the awards presentations.

Feedback reports provided — After November 23

The foundation is pleased to receive feedback from applicants at any time, but will specifically request it after the site visits have been completed. Feedback from applicants is essential for the continuous improvement of the award processes.

Feedback from applicants

Applicants are also encouraged to become actively involved in review and improvement of the application guidelines and associated evaluation criteria, and of the evaluation process.

Figure 18.2. *(continued).*

Important Points

Supplementary materials may be included with the detailed submission, but the following points should be considered.

- Only the detailed submission is evaluated; the evaluators are not bound to consider supplementary material. Such material should only be provided by way of example, (for example, a copy of the organization's staff newsletter to illustrate internal communication).

- All important data should be included in the text of the detailed submission. Do not include such data in the supplemental material and cross-reference it in the text.

- When writing the detailed submission, avoid using colloquialisms, acronyms, abbreviations, or other such terms to describe activities or approaches, even when such terms are widely used and understood within the applicant's business field. When such terms must be used to accurately describe the organization's activities they must be defined clearly when first used. Do not use such terms in place of a description of the activity in the belief that the evaluators will (or should) understand what is meant (see Figure 18.3).

The Site Visit

The decision on whether to conduct a site visit is made by the evaluation panel. A site visit should be planned for by all applicants, although it is not an automatic part of the assessment procedure. Applicants selected for follow-up site visits will be notified within a few days of the completion of the first stage of evaluation and must be prepared for a site visit by the evaluation team at some time between Monday, September 13 and Friday, September 24, 1993. The duration of the visit will be dependent on the size, complexity, and geographic spread of the organization. It is at the discretion of the evaluation panel and will usually take a minimum of one day.

Although every attempt will be made to meet the wishes of successful candidates concerning the timing of site visits and to adhere to the timetable, the Australian Quality Awards Foundation reserves the right to nominate the date of the site visit within or outside the time given.

The Evaluation Process
A detailed evaluation of all entries is made by an evaluation panel that reports on that assessment to the panel of judges. The evaluation panel operates under the chairmanship of John Lysaght, a leading businessman prior to his retirement. The panel comprises experienced managers and quality specialists, carefully selected for their capacity to participate expertly and effectively in the evaluation process.

The Judging Process
Judges, upon receipt of the evaluation panel's reports, and after such other investigations as they choose to undertake, make the final selection of award winners. The judging panel comprises a number of leading businessmen, academics, and government representatives.

The Feedback Report
At the conclusion of the evaluation and judging process a written report, incorporating the evaluators' and judges' assessment of the organization's performance, is forwarded to each applicant, including those not selected for a site visit.

An appropriate evaluation team, drawn from the evaluation panel and under the guidance of an experienced team leader, is assigned to the application.

The applicant's detailed submission is assessed by the members of the evaluation team individually. A report is prepared by each team member.

The team meets, reaches consensus on each application, and selects those entries for whom a site visit is considered appropriate.

The evaluation team meets on-site with those organizations selected to progress to that stage.

The evaluation team prepares a final report, including recommendations, which is forwarded to the panel of judges for consideration.

The panel of judges meets with representatives of the evaluation panel and team leaders, reviews each of the applications that received a site visit, and determines the winners of awards in each section.

Judges and members of the evaluation panel are not permitted to participate in the evaluation of an application if they have an interest in the organization involved or in the outcome of the evaluation. Applicants are provided with an opportunity to raise objections to members of the team evaluating their application.

Figure 18.3. Evaluation and judging procedure of the Australian Quality Award.

The purpose of the site visit is

- To clarify issues raised during the evaluation of the written application. It also provides an opportunity to update the data presented in the written application.

- To verify the major strengths of the application.

- To verify that the written application truly reflects the organization's systems and processes.

- To investigate areas that are difficult to describe (and understand) in a short document.

- To determine additional facts.

- To amend the consensus report that will form the basis of both the report to the panel of judges and the feedback report to the applicant.

- To make a recommendation to the panel of judges.

In the event that a site visit is recommended, the applicant will be provided with a booklet that provides more detail on how the visit will be conducted, the responsibilities of the evaluation team and the applicant in preparation for the visit, and what can be expected during the visit.

The Feedback Report

As soon as possible after the presentation of the 1993 awards a written report incorporating the evaluators' assessment of the organization's performance against each of the evaluation criteria will be forwarded from the Australian Quality Awards Foundation to each applicant.

The report will be comprehensive and will detail the assessment of the evaluators and judges on the strengths, and possible areas of opportunity, exhibited by the applicant against the criteria.

Chapter 19:

Quality Award Criteria Signpost the Way to Excellence in Quality Performance

TELE JUHANI ANTTILA

In high-level debates among European quality experts (within EOQ and EFQM), concern has been expressed about the over-emphasized concentration of European enterprises on technical and formal issues relating to quality systems. These aspects do not promote genuine, competitive quality performance. The trend should be toward a broader concept of TQM, for which national and international quality award criteria are recognized as an accepted framework.

Finnish Quality Award

The world's leading criteria for a quality award are those for the MBNQA. This set of criteria has been developed and matured over the years through experience and improvement. The criteria for the European Quality Award are expected to be an important factor in raising quality management in Europe to world level. However, the development of the European criteria is still in its early stages, and also in Europe at corporate level, the Baldrige Award principles are most often applied. In practice the two sets of criteria are very close to each other.

The Finnish Quality Award, which is now (1994) in its fourth year, is presented annually on World Quality Day by the Finnish prime minister. The award criteria have been updated year by year on the basis of the latest American and European experiences. The evaluation routines are also under continuous development to ensure that they correspond to international models. The companies applying for the award or using its principles in their internal quality development can, of course, make use of the abundant, detailed material which has been published on other recognized models for the Finnish award, in particular the Baldrige Award.

Standards Should Be Complemented by Quality Award Criteria

The ISO 9000 standards presuppose that enterprises competing in quality should apply not merely the standards, but also the self-assessment approach for which the quality award criteria are very suitable. Being based on consensus, the standards always represent mediocrity and are not always updated. For example, the ISO 9000-9004 standards, approved in 1987, actually represent the quality thinking of the 1970s. Even though the ISO 9000 standards are being continually developed, this development is very slow, while the quality award criteria are updated annually.

The standards emphasize system-related aspects while the quality award criteria focus on results achieved through systematic measures and leading to customer satisfaction. The TQM criteria for quality awards aim at excellence – competitiveness in quality – and they have been demonstrated to have significance for the success of the company in business. Of the ISO 9000 standards, ISO 9001-9003, which have been used in the third-party certification of quality systems, only cover the field of quality assurance.

Quality management and quality systems of the companies which are in accordance with the quality award criteria also naturally comply with ISO 9000; on the other hand, concentrating only on the certification of quality assurance systems has often impaired the overall development of customer-oriented quality activities. Quality system certification by a third party has even sometimes been felt to do more harm than good from the customer's point of view.

The Quality Award Criteria Embody an Internationally Recognized Assessment Scale Based on Business Principles

The subareas of TQM are weighted in the quality award criteria according to their significance for business operations. Customer focus and customer satisfaction thus rank as most important in the score. In assessing for the quality award, the methods used for quality, the extent of their application, and the results attained are considered.

The quality award criteria do not prescribe any specific way of going about quality procedures. The evaluation system emphasizes reasonable methods, the extent to which they are applied in business functions and their significance for business, and the positive trend and level of results in view of the competitive situation. The evaluation criteria can be tightened as the overall business situation changes.

Internal Quality Development Within the Organization

The winners of the Finnish Quality Award are concrete examples of companies that have successfully implemented TQM. The main objective of the quality award criteria is, however, to promote competitiveness in business on the organizational and on the national level.

Quality award evaluation brings out the strengths and weaknesses of the company's business operations, and the criteria are thus excellent tools for continuous quality improvement. They can be applied systematically to search out the points in the company's activities where attention is required.

Many companies have developed their own modifications on the basis of the quality award criteria. TQM evaluation is experienced as a positive, interesting, and important opportunity for learning, and it also provides a good framework for effective discussion and planning of development, in which top management also participates. For this reason, it is important that assessment of the company's quality status should be self-assessment, and it should be kept in the hands of the company's own management, not entrusted to consultants or certification bodies.

Evaluation Principles of the Finnish Quality Award

Awarding principles, weighted by category in 1992, are shown in Figure 19.1.

Assessment Criteria

When assessing the qualifications of a candidate for the Finnish Quality Award, the awarding committee estimates how well the company meets the following eight criteria of a quality firm (see Figure 19.2). By category, the following issues are considered:

- The company's systematic principles in dealing with requirements of each category

Category	Weight
1. Quality actions by senior management	70
2. Information and its analysis	70
3. Strategic quality planning	50
4. Human resources development and involvement	150
5. Quality management of operational processes	140
6. Quality results and operational results	170
7. Customer focus and customer satisfaction	300
8. Positive social impacts	50
Total	1000

Figure 19.1.   Scoring guidelines for the Finnish Quality Award.

- Implementation of the principles
- Operational results and improvement plans for the future

1. **Quality actions by senior management**

   The management of the candidate company is based on distinct and perceptible quality values established and maintained by its executives. Through their personal actions, the senior executives will lead the company in operations to achieve customer satisfaction as the end result.

2. **Information and its analysis**

   The management of the company is based on facts, ensuring customer satisfaction. The company has sufficiently extensive and updated information at its disposal, to be analyzed and made use of. Products and operations are compared with those of the leading companies.

3. **Strategic quality planning**

   Quality is a key factor in the planning process of the company. The objective of quality planning is to achieve and maintain a favorable competitive position on the market by combining quality improvement with the general business development. The assessment objects are the short- and long-term business plans and goals of the company.

4. **Human resources development and involvement**

   The company improves and makes efficient use of the competence of its entire personnel to meet the quality goals. The development of both the individual and the organization are considered in the assessment.

5. **Quality management of operational processes**

   The company maintains systematic procedures to continuously improve products and operations by reliable and controlled processes. These procedures also cover the purchase of materials, components, and services.

6. **Quality results and operational results**

   The company maintains a system of result parameters concerning its products and operations as well as improvements in these. The result figures are based on objective measurements, derived from customer requirements and expectations and business analysis. The results are compared with those of the competitors.

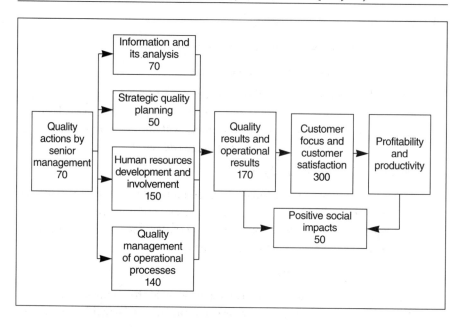

Figure 19.2. Evaluation principles for the Finnish Quality Award.

7. Customer focus and customer satisfaction

The company is informed of the requirements and expectations of its customers and is able to meet these with its products and services. The assessment criteria are customer satisfaction and its development as well as anticipation of the future requirements and expectations of the customers.

8. Positive social impacts

Through its operations and products, the company has a positive effect on the well-being of people, on the environment, and on natural resources. The company promotes general quality consciousness in Finnish society through its actions. An assessment criterion is the profitability of the company, because only a profitable firm can, in the long run, achieve positive effects on society.

Chapter 20:

The Israeli Quality Award

JUDAH L. LANDO   YOSEPH BESTER

In early 1990, the general assembly of Israel's AEI approved the initiation of an industrywide annual quality prize to encourage the adoption of the total quality philosophy by its members and to serve as a guide to others. The task of design, development, and implementation was assigned to the association's Quality and Reliability Forum. The charter explicitly stated that the goal was advancing and recognizing world-class quality.

AEI comprises essentially all of Israel's leading electronics companies. Member organizations design, develop, manufacture, sell, and service worldwide products ranging from components and subassemblies up to large-scale systems spanning a broad spectrum of industrial, commercial, and defense markets. A major player on Israel's industrial stage with sales totaling $3.6 billion in 1991 and employing some 33,000 people, the industry contributed significantly to the national balance of payments by exporting $2.3 billion to more than 100 countries across the globe. Led by a presidium of top CEOs, the association has consistently emphasized the need for its members to strive to be among the best world-class companies.

Strict customer demands for quality and reliability were imposed in the late 1960s as the then primarily defense-oriented industry labored to meet national needs after the Six Day War and subsequent embargoes imposed on Israel by previous sources. The emphasis on quality carried over into professional and consumer products as the industry and its member companies grew both domestically and in exports. By the late 1980s industry leaders recognized that meeting specifications and standards was not enough and that a quantum jump to total quality was prerequisite to achieving the business and national goals the industry set for itself.

In late 1989 the presidium decided to institute a quality prize competition whose goals were

- To publicize and advance the awareness of the importance of total quality

- To encourage companies and organizations to embark upon an ongoing improvement process

- To encourage member company cooperation as a lever to enhance international competitive position

- To bring to public attention, domestically and abroad, that the Israel electronics industry is actively striving toward excellence

- To encourage other sectors in Israel to initiate similar activities aimed at advancing quality behavior and to plot the path to quality by the prize criteria

- To provide the winners with recognition of their achievements as a worldwide marketing tool

A goal was thus set to design and implement a quality prize competition that would meet all of these criteria. The task of achieving this goal was assigned to the Quality and Reliability Forum of the association – a group comprised of senior quality, reliability, and productivity executives from the member companies. The presidium explicitly directed that the objective is world-class quality – no discounts or considerations allowed.

The target date for the first award was set as the next annual general assembly of the association (January 1991).

Process and Execution

The task was reported to the forum at its first meeting in 1990 and was willingly accepted as the first priority activity for the year. Execution was assigned to a working group comprising the leadership committee and the executive director of the association. Judah L. Lando, as forum chairman, was elected group leader.

The working group immediately convened and drew up a first draft Work Breakdown Structure (WBS), defined working procedures, and set staged target dates. The main elements of the WBS were

- Criteria definition and phrasing

- Competition rules

- Documentation preparation and dissemination

- Selection and recruitment of judges

- Assessment instructions and briefing judges

- Competitor submission review

- Assessment (judgment) execution

- Assessment report review, analysis, and preparation of recommendation report to presidium

- Ongoing reporting to the presidium subcommittee charged with overseeing the program

Members of the association received a package containing an application form, articles and rules, and an introductory cover letter. A nominal participation charge (the equivalent of $100) was levied to separate serious competitors from those who might misunderstand the magnitude of the challenge. Another device intended to make sure that the decision to compete was taken at the top level was to require the signature of the CEO on the application form. Forms signed by other personnel were returned.

Each assessment was carried out by a team of two judges. Assessment reports and scoring were forwarded to the working group and kept in strict confidence. To prevent concern that proprietary information might leak, group members who represented competitors of assessed companies stepped out during discussions of their assessment reports.

Analysis and Corrective Action

Analysis of the first year's assessments clearly showed that Pareto was on the job. The working group unanimously recommended to the presidium that Motorola (Israel) and National Semiconductor (Israel) be recognized as clear winners.

The recommendation was accepted and the awards were made by the president of the state of Israel, Chaim Herzog, at a gala dinner on January 3, 1991. But the analysis also showed that corrective action was needed to improve the process for the 1991 competition.

An in-depth review of the 1990 cycle was held by the working group and the following areas were targeted for improvement.

- The criteria and rules required revision, expansion, and preparation of a much more extensive guiding document.

- The load on the volunteer judges was close to the time they could reasonably be expected to contribute; even more so if the number of competitors in 1991 increased as was expected, hence more judges were needed.

- Uniformity between judges and assessments was deemed insufficient; judgment criteria and instruction of judges needed expansion and documentation.

- Increased participation was expected to lead to closer scores; hence more definitive differentiation was identified as a requirement.

- Smaller companies were perceived to be at a disadvantage when being compared with larger organizations.

- Recognition for achievements that fell short of being the best in class was warranted.

A review of the working group's operating experience also showed that, in all cases, decisions were taken unanimously. Consensus was the rule. This point is worth emphasizing. While each member of the working group was a senior executive and held strong views about how things should be done, after each point was deliberated, harmony was reached and unspoken agreement was reached that the shared appreciation of colleagues' wisdom would inevitably lead to unanimity. As a result of the high level of understanding and mutual respect reached, it proved possible to allocate tasks among subgroups with minimum review or revision required by the entire working group. That both expedited the process of the corrective action phase and greatly enhanced the productivity of the group as a whole.

It should also be noted that each member was able to contribute the use of his employer's facilities in such mundane aspects as typing, photocopying, and mail and fax services. Management support was manifested universally.

The corrective action implemented for the following year addressed each of the points listed earlier. New books written for the 1991 competition (articles and rules for the competition and judges' briefing) totaled almost 200 pages. A full-day briefing session attended by the entire panel and open to competitor representatives was held in October, shortly after which the assessments were conducted. It might be added here that an English language version of the articles and rules was published by the association and has already gone through three printings with extensive interest from companies in Israel and abroad.

Two competition categories were defined; small companies with up to 150 employees and larger companies.

Honorable mention awards were instituted for those companies whose quality systems and results manifested clear commitment and achievement but still had a way to go before they reached the top of the ladder.

After the first round of assessments was completed and the scores reviewed in considerable depth by the working group, a short list of candidates was drawn up and the final judging scheduled. Under the rules of the competition, winners are excluded from competing for two years. The final assessment team was comprised of the quality managers of the previous year's winners. Selection between the final few after the second round of assessment was also unanimous as were both the preparation of the recommendation to the presidium and the presidium's discussion. The decisions reached included awarding the 1991 quality award to Elbit Computers, and an honorable mention to Telrad and Tadiran Telecommunications in the large company sector and to Efrat among the smaller companies. The awards were again presented in impressive form at the annual gala dinner of the association.

The goals set in the original decision were achieved; the main objectives of creating an ongoing process and encouraging others to emulate this process are manifested in the current round and the intention announced by other organizations in Israel to initiate either their own competitions or to establish a national competition. The efforts invested by contestants and the practices they implemented are paying off daily. With a few exceptions, participants agreed that the steps they took to compete are in their benefit and essentially all of the companies that didn't win in the first two years are again trying. The winners widely publicized their awards and are all honorably discharging their commitment to share with others their methods and skills. Their openness alone in making their knowledge, procedures, and results available to almost anyone that asks for an opportunity to learn from them either by on-site visits or presentations by senior staff is highly laudable in itself.

During the first two years, many of the tools of TQM were used in the process and contributed to the success of the program. These included

- Vision by top management
- Clearly defined goals – overall and in detailed tasks
- Training – the working group itself, judges, and competitor representatives
- Documentation
- Execution according to plan and on schedule
- Review by top management - the presidium
- Candid self-assessment and corrective action
- Customer involvement and input

Vision, leadership, fair and open competition, and on unswerving commitment to quality all led by top management are vital to achieving world-class quality. Each of these elements is continually manifested by the association's leading members and serve as a real object lesson. The association's quality award catalyzed many companies to make these elements part of their culture. All that is left to be said is that the member companies of AEI are continuing to be an example that can and should be emulated by organizations in Israel and around the globe.

The Israeli Quality Award 1993 for Industry

In 1993, the Quality Award contest was opened to all Israeli industries and service suppliers. The 1993 Israeli Quality Award for Industry was presented by the prime minister of Israel, Yitzhak Rabin, to Motorola Israel. Motorola Israel received the award for its management commitment to the total quality approach and for its efforts for total customer satisfaction. Twenty-eight companies competed for the award. Nineteen were from large firms while nine were from small plants. This was the first time such a competition representing all of Israel's industrial market took place. The award selection was determined according to the Baldrige Award criteria adapted to the Israeli environment.

Section III:

Localized Quality Awards

Chapter 21:

Designing and Implementing a State Quality Award

Eric N. Dobson

Organizing a State Quality Award Effort

A state quality award can be initiated in various ways. Efforts can be started by the governor's office, by state policymakers, by private sector leaders, or by grassroots organizations. State quality award efforts are usually launched by both public and private sector individuals involved in state, regional, local, and/or private-sector quality initiatives and looking for ways to bolster the competitiveness of firms in the state. In the beginning stages, several independent groups may be working simultaneously to initiate a state quality award. These groups eventually form the backbone of the effort to establish a state quality award.

Identification of Interested Parties

Participation should be sought from individuals who are already involved in public and private quality initiatives. Other people can be recruited from groups interested in quality issues, such as productivity centers, chambers of commerce, and quality expert networks; state officials from the governor's office and economic development and labor departments; higher education interests; and representatives from industry, labor, and local government. These individuals will be important in coordinating the effort and in providing expertise and political influence toward creating a state award program.

Development of a Core Group of Quality Experts

The successful establishment and maintenance of a quality award relies on the integrity and credibility of the individuals supporting the effort. It is important that, as a state award is being planned, a core group of business executives involved with quality play a key role in providing guidance and input. Business executives involved with private sector quality efforts or participants in the Baldrige Award process can provide valuable expertise and contributions to a state quality award. Their involvement and input will strengthen the state program and provide a feeling of ownership of the award in the business community. The core group should also provide outreach to other businesses and the community in general about the value and benefits of an awards program. The involvement of business executives in a state quality award also conveys to the governor the level of interest and commitment to the award by the business community. When the state quality award is implemented, the business executives may be drawn upon to serve as judges or examiners for the award. In Texas a state initiative to promote quality and establish a state quality award is benefiting from the contributions of a loaned executive from Xerox and the expertise of Baldrige Award examiners.

Assignment of State Government Lead

Regardless of the level of state government involvement there is a clear need to designate a state government representative to serve as a key contact point and resource. The contact person may come from the governor's office, the state economic development office, or another state agency. This contact must be accessible to anyone seeking information on the state quality award. The contact person is important in demonstrating that a true public/private partnership exists for administering the award. In Wyoming, the state Department of Community and Economic Development runs the Governor's Quality Award with one key staff person who serves as a contact point for all inquiries regarding the award.

Identification of Existing Award Programs

Numerous awards already exist at the company and community level. Many large corporations (such as AT&T, General Electric, Texaco, UNISYS, and Westinghouse) have internal quality programs and quality awards. In four states, U.S. Senate Productivity Awards are presented annually as a means of promoting and recognizing productivity improvements. These existing programs are a good source of information about administering an award and about industry trends in the state. They are also a good place to look for individuals who may be interested in a statewide effort. It is important for award planners to learn about existing programs to ensure that the state quality award compliments existing local and private-sector efforts. The Minnesota Council for Quality works closely with communities to establish and maintain community quality councils. Currently there are 16 councils statewide. Made up of people from both the public and private sector, these councils are involved in numerous activities, including education and training, quality needs assessments for public and private organizations, public relations, and recognition awards.

Designing an Award Program

The design of a state quality award will affect the credibility of the award concept and indicate what issues are important to the state. An award needs to include a mission statement; eligibility guidelines concerning firm sizes and sectors; an application fee schedule; a list of the criteria by which applicants will be judged; information concerning the scoring of applications; a description of the application and evaluation processes; information on the selection, training, and role of examiners and judges; a description of the award presentation; and information concerning the obligations of winners. (See Table 21.1 for a summary of designs of existing state programs.)

Table 21.1.   Existing state quality award programs.

State	Award	Administered by	Year Estab.	Award Categories	Budget	Fees	Staff	Examiners/ Judges	1992 Applicants	1992 Site Visits
Delaware	State of Delaware Quality Award	Delaware Quality Consortium	1992	large and small: manufacturing and nonmanufacturing; nonprofit organizations	$102,400	more than 25 employees: $300; less than 25 employees: $100; nonprofits: $100	n/a	69	20	10
Florida	Governor's Sterling Award	Florida's Sterling Council	1992	small/medium (less than 200 employees for manufacturing and service; less than 100 for others) and large; private; public education and health sectors	$223,950	eligibility determination: $50; small/medium: $250; large: $500; site visits: $250 for small and $500 for large	2	n/a	n/a	n/a
Maine	Margaret Chase Smith Maine State Quality Award	Maine Chamber of Commerce and Industry	1989	large and small: manufacturing and service firms	$17,900	eligibility fee: none; eligibility review fee: large $100, small $50; advanced evaluation fee: large $200, small $100; site evaluation fee: large $400, small $200	1	9	4	0
Massachusetts	Massachusetts Quality Award	Massachusetts Council for Quality	1991	manufacturing, service, small business, and nonprofit	$157,100	eligibility fee: $50; manufacturing and service sectors: $150; small businesses and nonprofit: $650	1	40	28	8

State	Award	Administered by	Year Estab.	Award Categories	Budget	Fees	Staff	Examiners/ Judges	1992 Applicants	1992 Site Visits
Minnesota	Minnesota Quality Award	Minnesota Council for Quality	1990	manufacturing, service, and small business	$533,000*	eligibility: $50; small business: $700; other companies: $2000	1	180	23	7
New Mexico	New Mexico Quality Award	New Mexico Quality Foundation	1992	small, medium and large: manufacturing and service and government	n/a	n/a	n/a	n/a	n/a	n/a
New York	Governor's Excelsior Award	Department of Economic Development and Department of Labor	1991	large and small: private, public, and education organizations	$790,900	eligibility: $50; large manufacturing and service: $2000, all others: $500	1	65	24	9
North Carolina	North Carolina Quality Leadership Award	North Carolina Quality Leadership Foundation	1990	large, medium, and small manufacturing and service and nonprofit	$481,000	eligibility: $100; small business: $750; medium business: $2250; large companies: $3250	4	55	12	5
Wyoming	Governor's Quality Award	Division of Economic and Community Development	1986	one award	$5000	none	1	n/a	n/a	n/a

Baldrige Award Compatibility

Because the Baldrige Award is so well respected, it should be one of the first resources examined by state award planners. Existing state award programs include various components of the Baldrige Award. The Baldrige Award can provide a model for examiner training techniques, evaluation criteria, scoring methods, and evaluation processes. States can use the previous year's Baldrige Award training materials for examiner training. The Baldrige Award process can also serve as a benchmark for evaluating the processes of a state program. For example, how does the state program compare to the Baldrige Award in terms of the length of the application process or time allocated to providing feedback to applicants? Some states have found certain aspects of the Baldrige Award, such as scoring, to be applicable to their state and have modeled their program after that and other components.

Development of a Mission/Objective Statement

A state quality award can emphasize different objectives, so it is important to develop a mission statement for the award. This statement should

- Outline the reasons for creating the award

- Explain how the award will benefit the state and the business community

- List any specific industry sectors toward which the award is geared

- Discuss how the award has been designed to resemble other state quality awards, as well as the Baldrige Award

- Explain how the award will be administered

The mission statement will serve as a guide for individuals and organizations wishing to participate in the state quality award program. It should be accompanied by a business plan outlining the steps that will be taken to meet the various objectives.

Criteria

The application should explain clearly the basis on which the application will be evaluated. The examination categories/issues used by MBNQA include

- Leadership

- Information and analysis

- Strategic quality planning
- Human resource development and management
- Management of process quality
- Quality and operational results
- Customer focus and satisfaction

These criteria serve as a basis for evaluating the firm and include elements that are crucial in defining, implementing, and sustaining a quality effort. Leadership criteria are used to evaluate the senior management's success in fostering quality values in the firm. Information and analysis criteria are used to evaluate a firm's ability to collect and analyze data for purposes of improving quality practices. Strategic quality planning criteria evaluate the firm's effectiveness in integrating customer quality requirements in the business plan of the firm. The human resource development and management criteria refer to the firm's efforts to develop and realize the full potential of the workforce to pursue the firm's quality and performance objectives. Criteria concerning management of process quality examine the systematic process the company uses to pursue ever-higher quality and company performance. Quality and operational results criteria examine the firm's level of quality and trends in improving quality, operational performance, and supplier quality. Customer focus and satisfaction criteria evaluate the firm's relationships with customers and its knowledge of their needs; the criteria also evaluate the firm's understanding of the factors that are key in determining competitiveness.

The seven categories used in the Baldrige Award are a complete representation of a fully integrated quality system built on the basis of continuous improvement. Several states including Maine, Massachusetts, Minnesota, and North Carolina have adopted these criteria in their own quality award programs. Delaware uses similar categories and captures much of the same information. Wyoming does not have formal criteria but rather uses consensus and discussion to judge applicants. Some states make adjustments to the criteria; for example, New York's criteria emphasize the issues of human resource development and labor management relationships.

Scoring

The seven criteria outlined earlier are given a maximum number of points that can be achieved. Applicants for the Baldrige Award may receive a possible total of 1000 points, with the following breakdowns: 90 for leadership; 80 for information and analysis; 60 for strategic quality planning; 150 for human resource development and management; 140 for management of process quality; 180 for quality and operational results; and 300 for customer satisfaction. Maine and New York also use scoring systems that total 1000 points, but have different breakdowns for the categories. The scoring of the awards is an evolving process. Table 21.2 shows different scoring practices used in 1992.

Scoring is based on three issues: approach, deployment, and results. The approach is evaluated to see what tools are being utilized and their effectiveness. Is the approach preventive and is it integrated and systematic? Is the approach based on quantitative information, and what types of mechanisms are in place internally for feedback and evaluation? Deployment is evaluated on how well the approach is applied. Does deployment cover all functions and work for all internal processes, activities, employees, and facilities? Does it cover all services and products? Results are evaluated in terms of a company's success in addressing customer needs. Judges look at the rate and range of quality improvement, the importance of improvements to the business, the level of quality that has been achieved, and how the results compare with others in that industry, as well as those in other industries.

Table 21.2.   Examples of scoring practices.

Examples:	Baldrige Award	Maine	New York
Leadership	90	100	170
Information	80	60	40
Planning	60	90	40
Human resources	150	150	250
Products/process	140	150	100
Results	180	150	150
Customers	300	300	250

Eligibility

In determining who is eligible for the award, there are several issues to consider. The first is the nature of the business-industry mix in the state. Is the state base primarily manufacturing or service? What are the sizes of these firms? Is the industrial base made up primarily of state firms or out-of-state conglomerates with in-state locations? What level of participation does the state want from public and nonprofit organizations? Allowing awards to be made to nonprofit, governmental, and educational sectors promotes quality in government and demonstrates the interrelationship of the three sectors. New York State has separate awards for the private, public, and educational sectors. Minnesota has a category for service sector firms and Delaware has one for nonprofit organizations. Many states have modeled their eligibility criteria on the Baldrige Award and are finding it necessary to add a category for midsized firms. Maine divides its applicants into large (more than 100 employees) and small (less than 100 employees), Wyoming has only one category for all firms, and North Carolina is adding a category for medium firms after having used only small (less than 100 employees) and large (more than 100 employees) categories for two years.

Factors to consider when developing award categories include sectors and the size of the company or organization. Divisions within the industry sector may include service and manufacturing or private, public, and nonprofit. Within the categories pertaining to the organizational size of the applicant, the breakdown should be based on the demographics of firms in the state or any specific targeted sector. Small firms are usually those with less than 200 employees, medium firms have 200 to 500 employees, and large firms have more than 500 employees. The size of the firm is usually designated on the basis of in-state operations only. Delaware, Maine, and North Carolina all designate small firms as those with less than 100 employees. New York does the same except with regard to the government and education categories, in which a small entity is defined as one with less than 500 employees.

States need to consider two things when determining categories: Is it best to start small and grow as experience and credibility develops? Or is it best to start with recognition of entities of all sizes and sectors? Award designers also have to be careful to create enough categories to support the award but not so many that the value of the award is diluted. Excessive numbers or categories can lead to difficulties in implementing and administering the award. One step that is being taken in several states to address this issue is the inclusion of an honor roll category. This designation is for firms that are well on their way toward meeting the standards set by the award but are not yet fully integrated quality firms. The honor roll provides additional recognition without creating new award categories.

Eligibility Restrictions

The award categories indicate which firms are eligible to apply. However, there are usually additional restrictions. Generally, eligibility for a state quality award is limited to entities that operate in the state. In the case of multistage conglomerates or national companies, states need to decide which operations, if any, can be judged. This usually requires making a determination concerning how much of the firm's operations or quality efforts are within the state and without any support from other branches or from corporate headquarters. States also have the option of restricting the award program to firms that have a majority or all their operations in the state. In Massachusetts, any firm or subsidiary may apply for the award, while in Wyoming, eligibility is limited to firms with a majority of operations in the state. Previous award winners are usually restricted from participating for several years.

Fees

There are three specific fees that can be instituted as part of a state quality awards program. These include an eligibility determination fee, an application fee, and an on-site visit fee. The eligibility determination fee is usually small, ranging from nothing to $100. This fee, which covers some of the early administrative costs, generally ensures that applications are submitted only by serious contenders. In some cases this fee is returned for unsuccessful applicants. Applicants who are declared eligible then submit complete application reports. Application fees can range from $500 to more than $2000. Depending on the structure of the state award program, there are often different fees assessed for different sized firms. In New York, application fees for large manufacturing and service applicants are $2000 while all other organizations pay $500. In Maine, the fee for the site visit is set at $400 for large firms and $200 for small firms, while in New York the site visit fee for large organizations reflects the actual site visit expenses; for small organizations, a site visit fee is negotiated.

Application and Evaluation Process

The majority of state quality award programs follow the calendar cycle, beginning in January and ending the following January (see Figure 21.1). The process cycle includes up to three phases (see Figure 21.2). Before the first phase, applications are mailed out and a deadline for submission is set. In some states, such as Minnesota, a fairly short original application is reviewed by examiners before the applicant needs to fill out a more extensive application.

The first phase of the award process includes receiving the applications and assigning them to examiners for evaluation. Each application is evaluated by several examiners. During this phase, usually in April and May, each examiner independently scores the applications that he or she has been assigned. The scores are reviewed by a panel of judges, which decides which applicants should continue to the second phase. Applicants who do not continue to the second phase receive a feedback report from the examiners. Those who continue receive a status report on their applications.

The second phase, in June and July, involves a more detailed scrutiny of the applicants. This phase is led by a senior examiner who works with several other examiners to reach a consensus. The team submits to the judges a consensus score for each item as well as a total score. The judges review this information and determine which applicants will receive site visits. Applicants who do not receive site visits are provided with feedback reports on their applications.

The third phase of the evaluation generally occurs in August and September and involves site visits by a team consisting of examiners and at least one senior examiner. Applicants host the examiners at their facility and at any appropriate outlying branch or location examiners may wish to visit. Site visits may last from two days to a full week, depending on the organization's size and number of sites. After the site visit, a written report is prepared for the judges. The judges make the final decision on which applicants, if any, have met the quality standards established by the award and recommend recipients for the award.

Examiners/Judges

It is critical that examiners and judges be well qualified and well trained for their task. These individuals must meet a high ethical standard and have specific experience in and knowledge about quality programs. Criteria used to evaluate Baldrige Award examiners relate to breadth of knowledge about total quality, length and types of

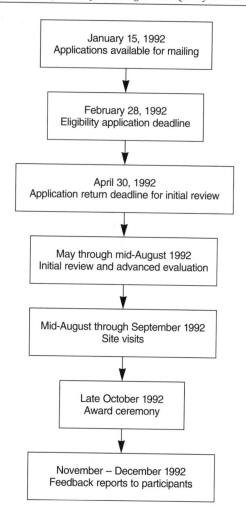

Figure 21.1. Award cycle for Maine State Quality Award – 1992 timeline and application review process.

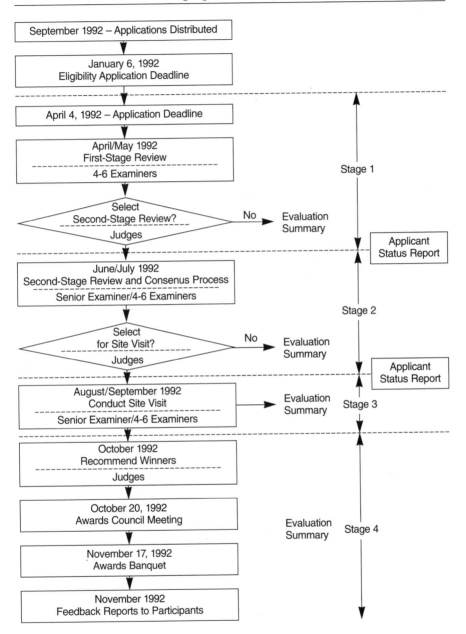

Figure 21.2. Award process for North Carolina Quality Leadership Award – 1992.

experience, communication skills, education and training, and achievements and recognition. There should also be adequate representation from the different sectors of industry to provide diversity among panel members. This diversity, coupled with specific standards on which applicants will be judged, will assist in providing fair and equitable judging across industry sectors.

Individuals interested in becoming examiners generally apply to the award administrator for consideration. The application for these positions requests information on work background, knowledge of and experiences with quality and quality practices, and agreement to an ethics statement. Unlike examiners, judges do not apply for their positions, but instead are sought by the award administrators for their particular expertise or knowledge. Many states use Baldrige Award examiners and judges in their own programs. The Baldrige Award officials not only serve as state examiners and judges, but also may conduct training for new examiners and judges.

There are generally three levels of examiners and judges. In most states there are examiners, senior examiners, and judges, with respectively more important responsibilities and experience. Using several different levels of review ensures accuracy as well as fairness in evaluations.

Most states use the Baldrige Award examiner training process to convey a full understanding of criteria and scoring, as well as the award process. Trainees learn how to utilize their knowledge of quality in evaluating the applications and how to apply overall concepts of quality across industry sectors. Examiners and judges must be willing to commit a significant amount of time to the award process – a total of approximately 10 to 20 days over a six-month period. Examiners are involved in all phases of the award process; their responsibilities include reviewing all applications, closely scrutinizing second-phase applicants, and evaluating firms during site visits. Senior examiners are involved in facilitating consensus, participating in site visits, and writing feedback reports.

The role of judges is to review information provided to them by examiners to ensure the credibility and accuracy of the information, and to make decisions based on their analysis. The judges carry considerable importance and are the most experienced of the quality experts participating in the application evaluation process.

Award Presentation

During the year of activity associated with a state quality award, excitement and interest tends to mount among participating firms and others. One way to capitalize on this interest is to present the annual award at a highly visible event. New York presents its award at a banquet hosted by the governor and lieutenant governor; North Carolina holds a three-day quality conference that includes presentation of the award; and Delaware holds a Sharing Rally in January, just before the next year's applications are made available.

Post-Award Expectations

The receipt of a state quality award is an important accomplishment for a firm and an excellent promotional opportunity for the award. The winner should be show-cased as an example of what the state is talking about when it refers to a quality firm. Winners of state quality awards serve as ambassadors of quality by sharing experiences. Many states require that the winning firm share its experiences by means of case studies, presentations, or site visits. These activities not only provide an opportunity to showcase the firms, but also serve as an opportunity to discuss the award and raise the general awareness of quality and its importance as a business strategy.

Implementing an Award Program

Once an award design has been completed, the task of implementing the award needs to begin. Implementation includes administering the award, determining the role of the public and private sectors, and securing funding. These steps can be undertaken in a variety of ways. The following discussion outlines the steps and some of the advantages and disadvantages to each. (See Figure 21.3 for an outline of the implementation of the Minnesota Quality Award.)

Establishment

A state quality award may be established in several different ways, and there are successful examples of each method. A state quality award can be established by executive order, providing the immediate leadership and support of the governor. With an executive order it is also important to maintain a broad base of support in the event of a gubernatorial transition. Several states have initiated quality awards as internally driven state staff functions. In such cases, the award is readily identified as a state

1986	1987		1988	
Governor's Commission on Minnesota's Economic Future established	Commission presents report on quality	Legislation and funding for quality council secured – $175,000	Initial research and programs (workshops, surveys of activity and interest)	Corps of volunteers developed

1989				
Staff hired	Private funds raised	Board of directors expanded	Statewide quality service award begun	Community councils coordinated

1990			1991			
Planning for quality award begun with $50,000	Membership expanded	First quality award presented	Second state quality award presented	Third quality service award	Quality conference held (650+ people)	Staff expanded

Figure 21.3.   Implementation of the Minnesota Quality Award.

government program, and many startup expenses are minimized. Another approach is for the state legislature to establish the award; this is possible if enough members support the initiative. A legislative statute ensures continued support through gubernatorial transitions and demonstrates political and financial legislative support.

Organizational Structure

There are a variety of ways to organize an awards program. The award can be administered by the state government, by a private sector foundation, or by a third party. Regardless of which framework is chosen, there are clear roles that the different sectors can play.

State Administered

In New York and Wyoming, an agency of the state government runs the state quality award. The agency handles all administrative activities of the award from announcements to award presentation. The advantage of this approach is that the award program can tap into existing staff and resources. Because of this, there may be less need for large up-front revenue to cover staff costs. In addition, the state is well positioned to tap into state, local, and regional human and financial resources from across the state. These resources can be valuable in marketing and administering the award.

When the state administers the quality award, consideration must be given to the impact of public disclosure laws on the applicants' confidentiality. Confidential company reports produced specifically for the award process, confidential internal company reports, and feedback reports provided to the company as part of its award evaluation could become public information. This risk can be mitigated by informing applicants that the application reports are confidential material loaned to the state for the purposes of evaluation, and that all material remains the property of the applicant and will be returned upon completion of the award evaluation. Steps also need to be taken to ensure that scores are maintained in such a way that specific applicants and/or examiners will not be disclosed.

A state-administered award needs to be structured in such a way that impacts from gubernatorial, legislative, and fiscal changes will be minimal. In addition, if the state administers the award, it is important that the state be perceived as an effectively managed organization and hence qualified to administer a quality award.

Private Nonprofit

North Carolina and Minnesota, among others, use a private nonprofit organization to run the state quality award program. In this approach a 501(c)3 organization is created to raise funds and administer the award. One advantage of this framework is that it is outside the political arena of the state. The 501(c)3 status is advantageous in that it encourages corporate donations by providing a tax deduction to contributing firms. Another advantage is that a private nonprofit group organized for the sole purpose of administering an award program can focus its energies exclusively on the award. In addition, with private sector involvement there will be a greater sense of ownership – and, thus, support and credibility – within the business community. In both North Carolina and Minnesota, a 501(c)3 organization focused on quality in the state runs the state quality award as one of several quality efforts. In North Carolina, designated employees of the foundation, excluding those involved in the award process, raise funds for the award.

If this particular route is chosen, there are a number of considerations. A private group may have to spend more energy gaining the governor's support and commitment to the award. Additionally, there will be a constant need to secure ongoing financing for the award from either corporations or the state. In addition, although the 501(c)3 status has benefits, it can be difficult and time-consuming to obtain.

Third-Party Administered

Another option is to contract the award administration to a third-party group, such as an association, university, or state chamber of commerce. In Massachusetts the state quality award is run by the University of Massachusetts, and in Maine the state chamber of commerce and industry administers the award. This type of administration appears to work well but may not carry the perception of strong state support. In addition, there are issues of accountability for both the state, because of any endorsement it may present, and the administrator, as well as other possible conflicts of interest with the third-party groups.

Business Plan

Establishment of a state quality award involves a significant amount of preparatory work. Decisions need to be made on issues ranging from how to fund the award to how to present it. These decisions should be made in advance and an implementation plan should be outlined in order to diminish possible confusion or complications. A business plan provides the framework for implementation. It should include

- Reasons for creating the award
- A list of potential sources of financial and administrative support
- A strategy for soliciting financial and human resources from state government and industry
- A description of the organizational structure that will administer the award
- Information concerning the nature and type of award criteria to be used
- An outline of the application and evaluation process
- An outline of the examiner selection and training process

- Information concerning the sectors and sizes of firms to be recognized

- A realistic timeline for implementation

A business plan with these items serves as the foundation for future decisions. It should be developed before any formal activities are undertaken.

Funding

Financing for the state quality award should be addressed early in the implementation process. There is a great deal of variance in how much money is needed to run a state quality award program. An independent public/private partnership will need to secure funding for startup staff time, marketing efforts, and application documents, while a state-run program may be able to more readily tap into existing funds and resources of state government. Budgets of existing state quality award programs range from $5000 plus in-kind staff support for Wyoming's state-administered program to $533,000 to run all the programs of the Minnesota Council for Quality, a private nonprofit group. Long-term funding commitments from both the public and private sectors are needed.

Current administrators suggest that a minimum of five years of funding be secured. Once an award is in operation, there will be limited additional revenue from applications and award-related products.

State Appropriation

The level of state involvement in the award's administration will affect the state's contribution. State funding for quality awards can range from minimal appropriations to significant budget outlays for staffings and administration. In North Carolina, the state provides $25,000 a year, which pays only for the actual award. Wyoming, by contrast, provides $5000, which is used solely for marketing the award. In New York, where the state administers the award, the state provides a direct grant of $35,000 to the award and contributes $500,000 worth of in-kind staff support. Given the fiscal constraints that states now face, these allocations may not be guaranteed.

Corporate Support

The primary purpose of the state quality award is to encourage firms to become more competitive through the adoption of quality initiatives. Larger firms can play a key role by encouraging suppliers and contractors to fully implement quality programs. These larger companies can be approached for corporate support. The 501(c)3 status of the quality award organization allows these corporate donations to be tax deductible. Awards generally offer several different levels of sponsorship. Memberships or other support can be solicited from smaller companies and private individuals who are interested in participating but cannot offer the financial resources to help endow the award. It is very important that the fund-raising be conducted separately from award administration to ensure the integrity and credibility of the award. In Delaware, corporations designated as Founders of the Delaware Quality Award provide a one-time contribution of $5000. In North Carolina, contributions to the North Carolina Quality Leadership Foundation range from memberships ($100 to $1600 per year) to corporate sponsorships ($5000 to $25,000 per year). In Minnesota, the current structure is such that only two companies sponsor the award each year.

In-Kind Donations

Many of the quality awards have used in-kind contributions to fill gaps during the early startup stages. The most common and needed contribution is in-kind staff support. This can include loaned executives familiar with quality management or administrative staff to support the quality award staff. In-kind donations are also made for office space, equipment, and examiners' time.

Award Process Fees

Once an award program is underway, fees for various processes serve as a solid revenue source. Fees in some instances are designed such that companies only pay them as they enter a new phase of the award process. States need to examine the legality and ethics of collecting fees for an award or program run by state government to ensure that no problems arise. It is also important to establish guidelines for the site visits to avoid excessive hospitality on the part of the applicant company.

Retailing of Products

As award programs become more established and respected in the state, there is an excellent opportunity to generate revenue from resources developed during previous year award cycles. One of the more successful products is a forum, such as a quality conference, during which the state quality award is presented. In addition, revenue can be generated by producing case studies of winning companies, providing examiner training material to companies looking for ways to train managers on quality issues, producing videos on quality and quality companies in the state, and presenting smaller regional conferences on quality. There is a large demand for specific information on quality and how firms can develop it. A state quality award program generates valuable information that can be disseminated throughout the state at a nominal fee.

A Long-Term Commitment to Quality

There are many reasons for beginning a state quality award and many different ways to design and implement the award. Whether the award is intended to encourage firms to adopt quality and productivity improvement strategies or to recognize firms that have instituted successful strategies and can serve as role models for others, a state quality award will benefit the state by creating a quality of life culture that will hopefully permeate the public and private sector. A quality award is an effective tool in promoting a state, promoting firms, and giving exposure to continued and improved competitiveness.

However, for a state quality award to be successful, top state leaders must be committed to it. The importance of this support cannot be overstated. The governor's support will draw attention to the importance being placed on quality, improved productivity, and competitiveness in the state. Furthermore, the state quality award needs the support of the legislature, which can provide additional exposure, input, and political support. In addition, it is critical that a state quality award have the backing of key private sector and business leaders in the state. The political, financial, and human resource support of corporate CEOs is crucial in conveying the message that the award is supported by the leaders of the business community. This leadership will provide the impetus for smaller firms to become involved in quality. This public/private partnership will serve to create a state quality award that meets the needs of both the state and the business sector.

It is important for award planners to be cognizant of any related activities taking place in the state. Support from community quality councils, quality groups, corporate quality advisors, and others is integral to establishing a state quality award. Early coordination with these groups will provide a framework for broad-reaching support and impact. The experiences of various quality experts involved with an award will prove valuable in designing, implementing, and promoting a state quality award. In addition, it is important that this wide-reaching effort be coordinated through one contact person who is accessible to both participants and interested parties.

As the award is designed, it is important to rely on the experiences of other states and award programs. There is no need to reinvent the wheel. The MBNQA and the existing state programs are all examples of successful strategies. States can use these programs as models and adjust them to their particular needs. Modeling a state quality award program after an existing one can help award planners to justify their decisions. In designing an award, planners should also build in a mechanism that will allow the award to change and improve over time in order to keep it current with developments in the fields of quality and productivity, as well as allowing it to be adjusted according to results of its own self-assessments.

Development of a state quality award is a long-term commitment. It is not a task that will be completed in several months. There are many factors to be considered before the award is publicized, ranging from tax considerations to personnel issues. Once the award is in place, long-term, ongoing financial, political, and human resource commitments are needed to sustain the program.

The establishment of a state quality award marks a new era in a state. It is an era in which the governor recognizes quality as an important part of invigorating the state and national economy. State government has a role to play in encouraging small and midsized firms to remain competitive. A state quality award gives both small and large firms a goal to reach and a standard to meet. The quality award serves as a roadmap for firms to assess, reassess, and improve operations and thereby become more competitive in the global economy. A state quality award focuses firms' attention on quality and recognizes those that meet a discriminating standard.

Chapter 22:

How New York Launched a State Quality Award in 15 Months

DAVID B. LUTHER

Quality management emerged as an exciting and potentially powerful force for statewide improvement with the initiation of New York's Excelsior Award. Named for the state's motto (*excelsior* means *ever upwards*), the award appropriately symbolizes the state's determination to drive forward the process of continuous improvement.

On May 21, 1992, governor Mario Cuomo presented Excelsior Award commemorations made of Steuben crystal to four recipients: two in the private sector, Motorola Automotive and Industrial Electronics Group Plant and Albany International Press Fabrics Division; one in the public sector, The New York State Police; and one to an educational organization, the Kenmore – Town of Tonawanda Union Free School District. In his remarks, Cuomo emphasized that quality applies to all parts of the economy and that the standard of living for citizens depends on achieving quality in the private sector, in the public sector, in education, and in all areas that support the citizens of the state.

For the winners, the awards ceremony and dinner was a major milestone in their quality journey. For many others – including state officials, union leaders, educators, CEOs, people from large and small businesses, and general supporters – it was the beginning of a new dimension in quality. For the 150-person team that developed and fielded the Excelsior Award, it was a major victory and proof of what a dedicated partnership could accomplish in 15 months.

An Overview of the Award Process

The Excelsior Award recognizes organizational excellence in the private, public, and education sectors. Organizations in each sector apply for the award by describing in writing how well they meet certain criteria. These criteria are based on those used by the Baldrige Award and reflect the same categories of leadership, information, planning, human resource excellence, quality processes, quality results, and customer satisfaction.

The application and review process is straightforward. Guidelines and applications are published and distributed. Organizations must apply by mid-November. Applications are reviewed first by individual examiners and then by teams of examiners in December and January. In the next step, completed by early March, a panel of judges selects the applications for companies that will move ahead in the process and receive a site visit by an examination team. Site visits allow examiners to clarify statements made in the written application and to verify that the application represents actual practice. Results of the site visit are returned to the judges, who make their final recommendations to the governor in May. The names of the award winners are made public at a ceremony later that month. In September, a conference provides a forum for the public to hear the winners' lessons and experiences.

One extremely important aspect of the award process is that every applicant receives a feedback report, prepared by examiners selected for their quality management expertise. The report describes in broad terms how the applicant fared in the evaluation. The report also comments on strengths and weaknesses of the application. This process provides valuable insight into an organization, and many consider it to be the best outcome of applying.

Developing the Excelsior Award: Experience Counts

Two enabling factors got the Excelsior Award team off to a running start. The first was the very existence of the Baldrige Award. Having a major national award in place for several years brings legitimacy to quality awards in general. The Baldrige Award design provided a model for many aspects of the Excelsior Award: the categories for examination; the concept of approach, deployment, and results; the notion of scoring bands that describe levels of performance; the 1000-point scoring system; and much of the examination process architecture.

The second enabling factor was the presence of individuals with substantial Baldrige Award experience on the Excelsior Award team. One former Baldrige Award judge and one Baldrige Award senior examiner provided leadership and knowledge that were invaluable in shaping the Excelsior Award. Five other Baldrige Award examiners participated as Excelsior Award examiners. Because of its experience, the Excelsior Award team could anticipate potential pitfalls, such issues as problems of variation in examiner scoring and the difficulties associated with producing timely, worthwhile feedback reports.

The real benefit of using the Baldrige Award model and drawing on the experiences of those involved in the Baldrige Award process, however, was the freedom it afforded to make innovative changes. Because the major parts were in place and the basics could be agreed on quickly, energies could be directed to tailoring the Excelsior Award to the unique needs of New York State.

The first change was dramatic and fundamental. It extended the concept of quality from private enterprise into two additional sectors: public agencies and educational organizations. This change reflected the high degree of interdependency between the sectors and the need for all to embrace quality management. Each new area, along with the private sector, was addressed individually. Although the basic concepts remained constant throughout all three, separate criteria guidelines and application requirements were developed for each sector, and a separate application manual was published for each.

The second change was the addition of the partnering concept to the award criteria. Partnering is described as the process by which an organization reaches out to create mutually beneficial alliances with employees, unions, customers, suppliers, communities, and others with whom it shares a common interest. This is especially important for employer-employee and organized labor partnerships, since a requirement of quality is to overcome adversarial relationships and build for the future on shared interests and mutual respect. Partnering was integrated into the leadership, human resources, and customer satisfaction categories.

The third change was the addition of diversity as an area for examination. This examination area asked applicants to describe what they are doing to ensure that all of the resources of the organization are extended to women and minorities. The social justice argument for diversity is obvious. There is also, however, a persuasive economic argument that says that survival in the future will depend on using all members of the workforce. The Excelsior Award asks for evidence that planning for and managing diversity is a requirement and is happening now.

A final change was the redistribution of points. More emphasis was placed on human resources, giving it a weight equal to customer satisfaction. (In the Baldrige Award, customer satisfaction is the category with the most available points.) This change stressed labor-management cooperation as a major requirement for achieving quality in the workplace. One lesson relearned and underscored through the Baldrige Award process is related to the old proverb that a chain is only as strong as its weakest link. Strengthening every link is the job of effective human resource management – to ensure that every individual in an organization is being developed and is able to make a full contribution to achieving the organization's goals.

The Key Partners

The Excelsior Award came to life because of the partnership of four major players, each with a unique role.

The first partner was the Excelsior Award Founders, which donated a total of $240,000 to pay for Excelsior I (the award's first cycle). The Excelsior Award Founders demonstrated an uncommon amount of faith. They donated $15,000 to $20,000 each to an unproven concept run by an untested team (see Figure 22.1).

The second Excelsior Award partner was the Executive Committee, a group of 18 volunteers and state agency executives who met monthly to design and deliver the Excelsior Award. This extraordinary group was made up of school officials, union leaders, businesspeople, and state agency officials. Many came from very high levels – CEOs and corporate officers, university deans, heads of agencies, high-level union leaders, and others (see Figure 22.2). The energy for the Excelsior Award came from the diversity of this committee and from the commitment of its individual members to come to consensus, accept responsibility, and perform reliably throughout the process. The Excelsior Award Executive Committee became a team in the best sense of the word: a group of people working toward a goal larger than any individual member could achieve alone.

The Excelsior Award Founders played a vital role in the establishment of the award process. Each donated $15,000 to $20,000 to establish it.

American Express Company
AT&T
Avis
Corning
Dale Carnegie & Associates
Delta Consulting Group
Dresser-Rand Company
Eastman Kodak Company
General Electric Company
Goulds Pumps
IBM Corporation
Metropolitan Life Insurance Company
New York Telephone Company
Niagara Mohawk Power Corporation
Xerox Corporation

Figure 22.1. The Excelsior Award Founders.

Organized Labor
Amalgamated Clothing and Textile Workers Union, Local 14 A
Civil Service Employees Association
Communications Workers of America, District No. 1
United Auto Workers, Local 424
International Union of Electronics, Local 313
New York State American Federation of Labor-Congress of Industrial Organizations
  (AFL-CIO)
New York State United Teachers
Service Employees International Union State Council
Teamsters Local 294, AFL-CIO

Public Sector
New York Sate Department of Labor
New York State Department of Economic Development
Governor's Office of Employee Relations
Office of the Lieutenant Governor
Education Sector
Hudson Valley Community College
Jamestown City Schools
Rochester Institute of Technology

Private Sector
AT&T
Avis
Corning
Dale Carnegie and Associates
Dresser-Rand Company
Eastman Kodak Company
Goulds Pumps
Quantum Performance Group
Xerox Corporation

Figure 22.2. The Excelsior Award Executive Committee: member organizations.

The third partner in the development of the Excelsior Award was the state itself, beginning at the top with Cuomo and lieutenant governor Stan Lundine and including numerous state agencies. The state partner provided three critical functions.

1. Championship. Cuomo first announced the formation of a statewide Quality at Work award in a powerful forum: his 1991 state-of-the-state message. He stressed the importance of a process similar to the Baldrige Award for furthering economic development and established a clear direction for

involvement by both labor and management. He called a press conference to announce the Excelsior Award competition when it became a reality and presented the first round of awards when the cycle was completed. The governor has since issued an executive order directing the implementation of quality management in the state.

Cuomo's other pivotal administrative decision was to appoint his lieutenant governor as the Excelsior Award project's champion. Lundine, as mayor of the small city of Jamestown, New York, some years prior to the establishment of the Baldrige Award, had shown himself to be committed to and knowledgeable about quality. Dismayed by the frustrations citizens were experiencing at the city treasurer's office, he announced that it should not take any longer to pay property taxes than it takes to get a hamburger at McDonald's, and then set about to make it happen. His contributions to developing the Excelsior Award process were both numerous and insightful. He listened to progress reports, encouraged support from labor and management, and provided help whenever needed. His aide is a permanent member of the Excelsior Award Executive Committee, and this ongoing communication and support has been essential for allowing the state agency staff to dedicate time and effort to the Excelsior Award.

2. Taking on an administrative load that is as essential as it is invisible. Preparing criteria booklets and getting them printed; accounting for expenditures; arranging facilities for meetings, training sessions, and conferences; responding to inquiries; arranging site visits; and acting as the first line of contact is critical to overall success. Doing it poorly would jeopardize the effort. Doing it well is expected. The state government did it well.

3. Participation as team members in the individual process groups just described. The staffs of the Department of Labor, the Department of Economic Development, and the Governor's Office of Employee Relations worked on teams, provided administrative support, and established a clear, timely, and informed channel to the state's political structure.

In addition to the Excelsior Award Founders, the Executive Committee, and the state itself, the fourth critical partner in developing the Excelsior Award was organized labor. Taking a position to support quality management is a tough assignment for today's union officials. There are countless examples of enlightened labor leadership that lost elections because of stands supporting quality. Labor in New York

State has fully supported the Excelsior Award effort. It is notable that, at its most recent convention, the New York State American Federation of Labor-Congress of Industrial Organizations (AFL-CIO) passed a resolution supporting quality management. Ed Cleary, head of the state's AFL-CIO, has been an active adviser and supporter of the Excelsior Award, and Ralph Catroppa, director of education and training, is co-chair of the Executive Committee. It is clear that, if labor support had not been there, the Excelsior Award could not have gone forward as successfully as it has and maybe not at all. Labor's guidance was especially valuable in avoiding the inadvertent phrases or actions that could potentially become major issues.

Identifying and Managing the Processes

The development and delivery of the Excelsior Award was in itself an exercise in quality management. Each process required for the success of the Excelsior Award was first identified and then managed by the Executive Committee. A team was assigned to each process, managed its own dates and milestones, and used feedback from various customers as the basis for continuous improvement. The real work was accomplished off-line, and monthly meetings were used to communicate results, define coordination opportunities, and deal with issues.

Each monthly five-hour meeting started and finished on time, a requirement for people with busy schedules. Each meeting closed with a review of open issues, assignment of due dates and individual responsibilities, and a process check to determine opportunities for improvement. Detailed minutes were essential in managing this complex set of tasks, as was an agreed-on agenda published before the meeting. Key tasks, dates, and interdependencies were also published with a detailed timeline that extended 12 to 18 months into the next cycle.

Ten processes were identified and managed.

1. **The application development process covered the creation, modification, and rationalization of the criteria.** A subcommittee was created for each of the three sectors, and an overall facilitator led the effort to ensure cross-functional rationality. The process team developed the criteria for each sector, addressed the need for additions or changes to the criteria, and adapted language to the needs of each sector.

    Since the requirements of each sector were not identical, this team also made sure that, where differences existed, they were there for a sound reason, not simply the result of variation among writers. Over time, as criteria became better understood, differences between sectors diminished.

2. **The application review process oversaw the logistics and activities required to get applications out to potential candidates, and then saw the completed applications through the examination process.** Many tasks are involved in this process.

Eligibility inquiries must be addressed according to carefully developed criteria, and responses must be sent. Applications must be assigned and sent to four to six examiners, who read and score them independently according to prescribed and weighted criteria. Examiner scores must be collected and loaded into a database that supports the judging process. From this information, the judges (a second tier of evaluators) determine which applications move to the second round of review. The examiners then arrive at a consensus score for each remaining applicant.

Excelsior Award judges review the consensus scores and the comments written by the examiners. During this process, they eliminate some applicants. The next task is for groups of examiners to make site visits to the remaining candidates to clarify all aspects of the written application and to verify that what is written accurately reflects actual practice. A state staff member is assigned to each site visit team to ensure consistency and to provide feedback for improving the process. The examiners produce another round of comments, and the judges review these to determine which applicants will be recommended to the governor as winners. Finally, all copies of the applications must be returned or destroyed to ensure confidentiality.

All these tasks required careful forethought and ongoing management at a level of considerable precision and detail. Information management is an example of one of the hidden but extremely important aspects of the application and review process. The Excelsior Award team did not start with dedicated hardware and software, but it quickly became evident that it needed to own the equipment and have complete control over data entry. Two computers were purchased and then programmed according to a complex set of end-product requirements. It was important both that the information be entered correctly and that it come out in a format that would be meaningful to the judges and efficient for them to use.

3. **The examiner and judge selection process was concerned with soliciting applications for the examiner and judge positions and for deciding who was selected.** The first round of the Excelsior Award used 67 examiners, and Excelsior II selected 110. Requests were solicited from all sectors,

and decision guidelines were established to make sure the proper balance was achieved. Criteria for selection included quality-related experience, managerial experience, and functional experience as a practitioner, academician, or consultant in a specific discipline such as human resources or QC. Diversity considerations played a role as well.

One early expectation was that a heterogeneous pool of examiners would all but self-destruct when they come together. Could examiners with different views of quality, different experiences, and different vocabularies train and work together effectively? Second, could examiners from one sector knowledgeably examine applications from another sector? To everyone's initial relief and subsequent delight, what was perceived as a potential problem became a strength. The training sessions quickly revealed that everyone learned from everyone else and that, once the basic concepts and language were clear, there was more commonality than was previously thought. Examiners were, in fact, assigned to sectors other than their own to create team balance, which also proved to be a strength.

4. **The examiner and judge training process was a major building block and success factor for the Excelsior Award.** The design and delivery of Excelsior Award training was led by a senior Baldrige Award examiner, Martin Mariner, and a former dean in a major university, Mark Blazey. The examiner training lasted for three days and was given four times in two locations in the state. The requirements for the training were rigorous. The training had to address all three sectors: public, private, and education. It had to review, and sometimes instill, the basic concepts of quality, and it had to describe the examination process. Finally, it had to teach team building and the process of reaching consensus – skills generally thought to be common, but realistically are fairly rare.

5. **The judging process, although invisible to the applicants, was absolutely essential to the evaluation and award process.** The key was in the selection of the judges themselves. The design called for nine judges who represented the Excelsior Award's various constituencies. All needed to have demonstrated substantial experience in their chosen fields, and be able to withstand potential public scrutiny. Judge candidates were selected by the co-chairman of the Executive Committee and recommended to the governor, who made the appointments.

The critical attribute of judges was to be able to go beyond bestowing awards to the organizations with the highest score. Judges had to determine which organizations scored well and best represented the purpose and ideals of the Excelsior Award.

In terms of the process, the identity of the applicants was not revealed to the judges until after site visit selections were made. Prior to that, all applicants were identified by code only. Applicants were ranked by score in descending order, and decisions were made starting with the lowest scoring applicant. Once an applicant was determined eligible to continue, all applicants with higher scores automatically made the same cut.

Several factors that might not be obvious had a major impact on the efficiency of the judging process. First was the format of the examination data. The design of information display had considerable impact on the ease with which judges could compare applicants. The second factor was accessible information about the examiners themselves. While consistent scoring was the object, round one saw considerable variations. Thus, it became important for judges to take into consideration the specific, relevant knowledge and expertise with which any given examiner approached a particular application.

6. **The feedback process provided every applicant, regardless of score, with a report describing the application's strengths, weaknesses, and general scoring position.** Actual scores were not given, but applicants were told of their positions within a range of 100 points. An initial feedback report was generated by the site visit team or other examiners in cases where there was no site visit. This was then given to the feedback team, which edited, published, and sent the final document to the applicant.

One problem with feedback was timing. The last feedback reports went out in July 1992; for some companies, that was eight months after they had applied. This caused a great deal of concern on the part of the applicants and, as they rightly pointed out, essentially prevented them from using the feedback as the basis for improvement for the following year's application. Questionnaires were sent to all applicants asking for comments and suggestions for improvement; they said timing was their biggest concern. For Excelsior II, the target was to reduce the elapsed time between application and receipt of report to 90 days. Barbara Harms at the New York State Department of Labor, was charged to ensure that this target and others were met. In fact, her energy and enthusiasm have helped to drive the Excelsior Award effort.

7. **Basic marketing – creating a general awareness of the Excelsior Award, quality, and their importance – constituted a major challenge.** For New Yorkers familiar with the Baldrige Award, the task was one of simple comparison. For others, however, the task was more difficult since the subject of quality, as defined by the Excelsior Award, was and is still very new – especially in the education and public sectors.

The announcement of the award by the governor was an important first step. Press releases were issued periodically during the first cycle of the award, and presentations were made in several cities by members of the state agencies.

Cuomo's executive order directing the implementation of quality in state agencies unquestionably helped the cause within the public sector. An Excelsior II media tour of the state was planned with the objective of raising awareness among the state's newspapers and radio and television stations. The real challenge, however, will be to convince all those who are eligible that continuous improvement is necessary for survival and that using the Excelsior Award application as a guide to improvement is a sound strategy.

8. **Events had relevance to marketing, training, and recognition, but were in themselves worthy of dedicated processes and effort.** There were two important events for Excelsior I. The first was the awards dinner, during which the governor presented the awards to the winners. This 500-person event was one of the highlights of the process and provided a great deal of recognition for all involved. The importance of the awards ceremony was emphasized by members of the Kenmore – Town of Tonawanda Union Free School District, who made the 600-mile round-trip from the Buffalo area to Albany and back the same night by bus.

The second event was a Winners Conference that gave the winners a public platform to describe their processes. In addition, a clinic was offered that described why and how to use the Excelsior Award application. These sessions were attended by some 400 prospective applicants from all sectors.

9. **Fund-raising efforts made the Excelsior Award a high-quality reality.** A small team of private sector members from the Executive Committee used personal contacts to make an appeal by letter, generally to the CEO of each of the Excelsior Award Founders. Ultimately, 90 percent of those contacted donated. A second campaign has just been completed with most of the same companies signing up for a three-year commitment ranging from $15,000 to $45,000. A more broadly based fund-raising effort is now being mounted to appeal to other large and small groups throughout the state.

The first round out-of-pocket cost was $225,000 and round two will climb to $300,000. However, this is less than half the real cost of operating the Excelsior Award. The all-important contribution of state agency staff makes it possible to run the Excelsior Award for this price.

10. **Financial planning and management became an ongoing process.** Setting up a chart of accounts, budgeting, managing costs, forecasting, accounting by cycle year, and reporting financial results to the partnership are all major activities.

Today's Issues

The issues facing the Excelsior Award today have changed substantially from those of just one year ago. The most significant change is one that the Baldrige Award team has debated for several years: qualifications versus competition. The Excelsior Award team has decided to move from a competition to a qualification approach in selecting winners. Excelsior I and II were structured to allow a maximum potential of two winners in each sector. Applicants were necessarily competing against each other.

Starting with Excelsior III, applicants must demonstrate a prescribed high standard of competence to the judges. That level will be consistent with past winners' levels. Every applicant that exceeds that level will be recognized, not just the current maximum of two. Hence, each applicant will be competing against itself rather than against others.

The second issue the Excelsior Award is addressing is the addition of a fourth sector. The first concern is whether the new sector will be healthcare or the not-for-profit sector (which could include some, but not all, healthcare institutions). A vote a few months ago on the issue resulted in not going after either at this time. That vote reflected the second concern: process capability. The team decided that it could not add a new sector at this time, because it was not ready. Discussion and analysis are continuing, and there most likely will be an additional sector added in the next couple of years.

The chairpersons are responsible for two more issues. One is continued fundraising. The challenge is to achieve a more broadly based funding support group. The second issue is renewal of the Executive Committee. Most members lasted for two cycles of the Excelsior Award, and some will last for three or more. The challenge is to select, recruit, and orient new members so effectively that continuity and progress of the process are ensured.

The final major issue is one of structure. A separate not-for-profit entity for the Excelsior Award is needed, with its own legal and financial structure, its own financing capability, and its own governance. The issue is to guarantee permanence and independence, and to make the Excelsior Award impervious to the possibility of a new, and potentially indifferent, state government. This issue is being addressed.

A Promising Future

The vision statement for the Excelsior Award asks that all of the institutions and organizations vital to New York State citizens' standard of living seek continuous improvement. The mission statement asks that a system of assessment be deployed that will help those institutions and organizations identify and implement quality improvement strategies.

Thanks to a team of more than 150 dedicated, hardworking people, 15 sponsors that are providing funding, strong state government and labor support, and a reliance on quality principles and processes, the first steps toward that vision and mission have been taken. The future of the Excelsior Award – and therefore, quality – in New York State is every bit as rich with promise as the beginnings have been.

Baldrige Award Provides Assistance with State and Foreign Quality Award Efforts

To further the intent of the MBNQA (to promote awareness of quality as an increasingly important element in competitiveness, understanding of the requirements for quality excellence, and sharing of information on successful quality strategies and the benefits derived from implementation of these strategies), the Office of Quality Programs at NIST works closely with states wishing to develop a state quality award.

Robert E. Chapman and Ann T. Rothgeb in the Office of Quality Programs interact regularly with these states, providing assistance and guidance in their efforts. Annually a workshop is held to share in information on Baldrige Award processes and provide networking opportunities with states that also willingly share information on how to design and manage a state award. The office maintains a listing of state quality awards that is provided to the public upon request and is regularly updated as new information is provided.

Jointly developed by the National Governor's Association and the Office of Quality Programs, a report entitled *Designing and Implementing a State Quality Award* is provided to states wishing assistance in getting started. In addition, the National Governors' Association has written *Promoting Quality Businesses, A State Action Agenda.*

The Office of Quality Programs also has received many requests from foreign countries for guidance in developing national quality awards in their countries. Various staff members have met with foreign representatives and shared Baldrige Award process information. Many multinational firms headquartered in the United States are also interested in foreign quality awards. A number of U.S. companies have subsidiaries in foreign countries and want these entities to become involved in that country's award process and to apply whenever possible. A number of foreign countries have now developed national quality awards that are patterned after the Baldrige Award. Modifications do occur to more closely meet the needs of the individual country. A list of the foreign country awards and their status is also maintained and provided to the public upon request.

Appendix A:

State Quality Awards in Place
(This partial list is provided as a service, for information only)

Alabama

Contact

Linda Vincent
Senate Award Coordinator
Alabama Productivity Center
Room 104, Farrah Hall
P.O. Box 870318
Tuscaloosa, AL 354847-0318
T – 205-348-8956
F – 205-348-9391

Status

8/93 (Information date): The award was created in 1986 and is patterned after the Baldrige Award. There are four award categories: small manufacturing (250 employees or less), large manufacturing (more than 250 employees), service sector, and award excellence in continuous productivity and quality improvement – previous winners. Certificates of merit are presented to the semifinalists for recognition of their achievements in productivity and quality improvement. Plaques are presented to the finalists (those who receive site visits) to recognize their excellence in productivity and quality improvement. Plaques and gold medallions are presented to the recipient(s) in each category (providing that applicants in a category qualify, meet the standards of evaluation, and are deemed worthy of recognition as leaders in increased productivity and excellence in quality). There is no limit to the amount of awards that may be given out per year.

Arizona

Contact

Dennis Sowards
Executive Director
Arizona Quality Alliance
1221 East Osborn, Suite 100
Phoenix, AZ 85014
T – 602-265-6141
F – 602-265-1262

Status

4/23/93 (Information date): Patterned much after the Baldrige Award, the Pioneer Award will be given only one time – in 1993 – and will be less rigorous than the 1994 award. The Pioneer Award will emphasize efforts more than results. All sectors will compete in the same categories. These categories are small (1-99 employees), medium (100-499 employees), and large (500 or more employees). For 1993 there will be no limit on the number of winners. These will include all sectors (private, public, nonprofit, education, state, but not federal government). In 1993 there will be CEO interviews, but no site visits are planned for the 1994 cycle. The 1993 Baldrige Award criteria will be used in 1993. Recipients will receive their awards at the Governor's Quality Conference on October 6, 1993.

California

Contact

Lance Barnett
Chief of Staff
Department of Consumer Affairs
400 R Street, Suite 3000
Sacramento, CA 95814
T – 916-445-1591

Robert N. Bridge
Executive Director
California Center for Quality, Education and Development
455 Capitol Mall, Suite 500
Sacramento, CA 95814
T – 916-321-5484
F – 916-448-2118

Status

3/10/94 (Information date): The Governor's Golden State Quality Awards is a joint project of the Trade and Commerce Agency and the Department of Consumer Affairs and has the support of the governor. The award is administered by the California Center for Quality, Education, and Development, a private nonprofit corporation organized under the laws of California, whose directors will be appointed by the governor. Hands-on administration of the awards program will rest with the Golden State Quality Awards Council, whose members will be appointed by the center's board. There are five award categories: (1) Quality in Management – recognizes companies that excel in managing their core activity, be it manufacturing, services, or a combination, (2) Quality in the Marketplace – recognizes companies that have translated an approach to quality into improved customer satisfaction and business results, (3) Quality in the Workplace – recognizes companies that realize the full potential of the workforce to achieve the company's performance objectives, (4) Quality in the Community – recognizes companies that excel in their interaction with their communities, and (5) The Governor's Golden State Quality Award – recognizes companies that best exemplify the qualities described in the award application criteria for all of the individual awards. California's program embraces the Baldrige Award's concepts, but it presents and expresses them so as to appeal to the broadest possible spectrum of business enterprises.

Contact
Thomas D. Hinton
President California Council for Quality and Service
7676 Hazard Center Drive, Suite 500
San Diego, CA 92108
T – 619-497-2599
F – 619-582-3709

Status

7/93 (Information date): The California Council for Quality and Service (CCQS) is a nonprofit, tax-exempt membership organization incorporated in 1992. CCQS administers the Eureka Award for California-based companies and their employees. The Eureka Award, a statewide quality award based on the MBNQA criteria, recognizes California-based service companies, nonprofit organizations, government entities, and educational institutions that have achieved quality excellence. The Eureka Award has five award levels.

Connecticut

Contact

Ernest J. Wagler
Project Facilitator
Connecticut Award for Excellence
P.O. Box 67
Rocky Hill, CT 06067
T – 800-392-2122
F – 203-258-4359

Status

7/18/93 (Information date): The Connecticut Award for Excellence (CAFE) is modeled after the Baldrige Award. The eligibility sectors are: (1) manufacturing – private sector, (2) service – private sector, (3) education – including public and private schools, colleges, and universities, (4) healthcare – including public and private organizations, and (5) public – including state and local governments. Within each sector, the award categories are subdivided into small-medium (300 or less employees) and large (300+ employees) organizations. Presently the award is being managed by staff of the Connecticut Department of Administrative Services' Personnel Development Center and supported by volunteers from business, healthcare, education, government, and labor.

Contact

Sheila Carmine
Director
Connecticut Quality Improvement Award
P.O. Box 1396
Stamford, CT 06904-1396
T – 203-322-9534
F – 203-329-2465

Status

1993 (Information date): Operated by the private sector, this program was launched in 1987. From its inception it has used the current year's MBNQA criteria for its applications. Award categories are small (1-100 employees), medium (101-500 employees), and large (more than 500 employees) for both service and manufacturing profit center facilities in Connecticut. One award may be given in each category for a total of six possible awards. Award recipients also are expected to participate as members of the Connecticut Quality Improvement Award Advisory Group for the following year.

Delaware

Contact

Michael J. Hare
Delaware Quality Consortium
Delaware Development Office
99 Kings Highway
P.O. Box 1401
Dover, DE 19903
T – 302-739-4271
F – 302-739-5749

Status

8/93 (Information date): The Delaware Quality Consortium, a not-for-profit foundation, was established to administer the award. The consortium is housed in the Delaware Development Office. There are a total of 10 possible awards per year, although there is no obligation to award all 10 awards., The award categories are manufacturing (large, small – two awards each), nonmanufacturing (large, small – two awards each), and not-for-profit (one size category – two possible awards). The award is based on the Baldrige Award, but uses a modified list of criteria categories. Formal announcement of the establishment of the award was made by governor Castle on March 27, 1992. First awards were presented in October 1992 and a sharing rally was held in January 1993.

Florida

Contact

John Pieno
Chairman
Florida Sterling Council
Governor's Sterling Award Office
Room 313, Carlton Building
Tallahassee, FL 32399-0001
T – 904-922-5316
F – 904-488-9578

Status

8/93 (Information date): The Governor's Sterling Award was established in 1992. It began as a regional award. The governor has now made it a state award. The award is administered by the Florida Sterling Council, a 501c(6) organization, which was established by proclamation of the governor. The council is chaired by the governor's executive director of the Florida Quality Initiative. There are no limits on the number of awards to be given. Based on the Baldrige Award criteria, it additionally includes financial and productivity sections. The categories are private manufacturing, private service, education, healthcare, and public. The first award ceremony and statewide quality award showcase were held in May 1993.

Louisiana

Contact

Ed O'Boyle
Louisiana Technical University
Office of Economics and Finance
P.O. Box 10318
Ruston, LA 71272
T – 318-257-3701
F – 318-257-4253

Status

7/29/93 (Information date): The Louisiana Senate Productivity Award has no budget. The award process began in 1984. One award can be given a year, private establishments only. Companies can nominate others and invitations to apply are mailed out through the senator's office. In 1988, the state began a second award for marketplace innovation with five major factors. The selection board is from winning CEOs – plant managers.

Maine

Contact

Nancy Werner
Maine Quality Center
Margaret Chase Smith Library
Norridgewock Avenue
P.O. Box 3152
Skowhegan, ME 04976
T or F – 207-685-3004 (call before sending fax)

Status

8/4/93 (Information date): The awards are modeled after the MBNQA. Any for-profit or nonprofit organization or appropriate subsidiary may apply. Awards will be considered in four categories: large manufacturers and large service organizations (100 or more employees), small manufacturers and small service organizations (fewer than 100 employees).

Maryland

Contact

Amit Gupta
Maryland Center for Quality and Productivity
College of Business & Management
University of Maryland
College Park, MD 20742-7215
T – 301-405-7099
F – 301-314-9119

Status

8/93 (Information date): Maryland does not have a separate quality award, but uses the Senate Productivity Award with criteria that address quality issues. The Maryland Productivity Award has been active since 1983. Senators Sarbanes and Makulski are very strong supporters. The program uses volunteer judges. There are three categories: manufacturing, service, and small business (100 employees or less).

Massachusetts

Contact

Brendon D. Healey
Executive Director
Massachusetts Council for Quality
92 Vernon Road
Scituate, MA 02066
T – 617-545-6200
F – 617-545-6200

Status

8/93 (Information date): The Massachusetts Quality Award is modeled after the MBNQA, but generally requires less detailed responses to questions. The award is administered by the Massachusetts Council for Quality, a nonprofit organization affiliated with the University of Massachusetts at Lowell. The categories include service, manufacturing, small business (200 full-time employees or less), and nonprofit (government, educational, healthcare, social service, and other nonprofit). The first award was presented at the Massachusetts Council for Quality Award Ceremony in October 1992. A separate award winners conference was held in June 1993. The award is named in honor of Armand V. Feigenbaum.

Michigan

Contact

Bill Kalmar
Director
Michigan Quality Council
Oakland University
101 North Foundation Hall
Rochester, MI 48309
T – 313-370-4552
F – 313-370-4462

Status

3/15/94 (Information date): The governor has issued a press release announcing the establishment of the Michigan Quality Council. The Michigan Quality Leadership Award will be based on the Baldrige Award. The Michigan award categories will include manufacturing, service, healthcare, education, public sector, and small enterprise. First awards are scheduled to be presented by the governor in the fourth quarter of 1994. If an organization receives an award, that organization and all of its subsidiaries are ineligible to win the Michigan Quality Leadership Award for a period of three years. However, recipients may submit an application in a following year for purposes of receiving a feedback report as part of their quest for continuous improvement. If a subsidiary receives an award, it is ineligible to win another award for a period of three years, but other subsidiaries of the same parent organization may be designated recipients. All may apply each year for feedback purposes.

Minnesota

Contact

Carol Gabor
Director
Minnesota Quality Award
Minnesota Council for Quality
3850 Metro Drive, Suite 633
Bloomington, MN 55425
T – 612-851-3181
F – 612-851-3183

Status

11/92 (Information date): The Minnesota Quality Award is sponsored and administered by the Minnesota Council for Quality. It is based on the Baldrige Award criteria, but does not include the areas to address. Award categories are manufacturing, service, and small business. An additional award category for education will be established in 1994. It is intended that categories for government and other nonprofit organizations be established at a later date. Please contact the Minnesota Quality Award director if you are interested in participating in a pilot assessment program. Up to two Highest Achievement Awards may be given each year in each of the three award categories.

Missouri

Contact

John Politi
Executive Director
Missouri Quality Award
P.O. Box 1709
411 Jefferson Street
Jefferson City, MO 65102
T – 314-634-2246
F – 314-634-4406

Status

2/94 (Information date): The award is patterned after the MBNQA. On June 29, 1992, governor Ashcroft signed an executive order creating the Missouri Quality Award. The award is administered by the Excellence in Missouri Foundation, a privately funded, not-for-profit organization. Award categories for 1994 are manufacturing and service, healthcare, education, and government are planned to be added. A maximum of two awards can be presented in each size class of the eligibility categories. Within each of the eligibility categories, awards may be presented in the following size classes by number of employees: zero to 99 employees, 100-499 employees, and more than 500 employees. Modest fees will be collected to cover some of the examination and site visit costs. The first awards are to be presented in October 1993.

Nebraska

Contact

The Edgerton Quality Award Program
Nebraska Department of Economic Development
Existing Business Assistance Division
P.O. Box 94666, 301 Centennial Mall South
Lincoln, NE 68509-4666
T – 402-471-4167 or 800-426-6505
F – 402-471-3778

Status

3/25/94 (Information date): The award process is patterned after the MBNQA process and Minnesota's Quality Award. Awards for continuous process improvement and adaptation of technology will be presented in both the manufacturing and in the service categories. Applicants will be required to select one category for examiners to review. The manufacturing category recognizes companies that have done an outstanding job of analyzing their manufacturing processes and initiating programs to continuously improve quality. A separate award will be presented to a manufacturer that has done an outstanding job in adapting technology to increase quality. The service category recognizes service companies that implemented quality improvement programs. As in the manufacturing category, a separate award will be provided to a company that has done an exceptional job of adapting technology. An eighth category in the application is "Sharing of Information" (5 percent weight). If the applicant wins the award, it must address how it plans to share information with other Nebraska firms.

Nevada

Contact

Katrina Ekedahl
Administrative Manager
Quality and Productivity Institute
P.O. Box 93416
Las Vegas, NV 89193-3416
T – 702-798-7292
F – 702-798-8653

Status

2/2/95 (Information date): Nevada was one of the early states to offer a U.S. Senate Productivity Award. It was originally modeled after the Virginia Award and in 1991 began using the Baldrige Award criteria as a guideline. All Nevada organizations are eligible. Nevada subdivisions of out-of-state organizations may apply as long as the quality and productivity initiatives originate and are implemented within Nevada. For 1995, categories for recognition include, but are not limited to, education, government, health and medical services, manufacturing and mining, nonprofit, resort and travel, and service and retail. The review of an applicant's initiatives addresses the same seven categories as the Baldrige Award.

New Jersey

Contact

Ed Nelson
Executive Director – Operations
New Jersey Quality Achievement Award
Mary G. Roebling Building, CN 827
Trenton, NJ 08625-0827
T – 609-777-0939
F – 609-777-2798

Status

8/2/93 (Information date): Prompted by governor Florio's Executive Order, the New Jersey Department of Commerce and Economic Development and the private volunteer organization, Quality New Jersey, have formed a public-private partnership to administer the award process. The New Jersey Quality Achievement Award is patterned after the Baldrige Award. Award categories are manufacturing, service, small business, education, and government. The first year of the awards was in 1993.

New Mexico

Contact

Julia Gabaldon
Executive Director
Quality New Mexico
320 Gold, S.W., Suite 1218
Albuquerque, NM 87102
T – 505-242-7903
F – 505-242-7940

Michael D. Silva
Award Administrator
Quality New Mexico
320 Gold, S.W., Suite 1218
Albuquerque, NM 87102
T – 505-242-7903
F – 505-242-7940

Status

4/18/94 (Information date): The New Mexico Quality Award is sponsored and administered by Quality New Mexico. The awards program has a tiered award structure. Level 1, Pinon Award (Commitment), requires a description of how the organization is approaching the development of an organizational quality system and any results achieved. Applicants must address each of the seven Baldrige Award categories. Level 2, Roadrunner Award (Progress), requires a description of the company's quality system development, implementation, and results achieved. Applicants address each of the items (28) in the Baldrige Award criteria. Level 3, Zia Award (Excellence), requires the description of the business and executive commitment, plus a description of how the company is applying all of the Baldrige Award criteria. These applications should be written to the area level of the criteria.

Trend data and results achieved must be presented. Any for-profit or nonprofit business or appropriate subsidiary located in the state of New Mexico may apply for an award. The award categories are: private sector manufacturing, private sector nonmanufacturing, and public sector: federal, state, and local. Currently, education is not eligible for an award (it will be next year) but may apply and will receive feedback. A small business is defined as one with 25 or fewer employees. Judges will make the final determinations for all awards.

New York

Contact

Barbara Ann Harms
New York State Department of Labor
Office of Labor Management Affairs
Harriman State Office Campus
Building 23, Room 540A
Albany, NY 12240
T – 518-457-6743
F – 518-457-0620

Joanne Fitz Gibbon
New York State Department of Economic Development
One Commerce Plaza
Albany, NY 12245
T – 518-474-1131
F – 518-474-1512

Status:

8/93 (Information date): The governor's Excelsior Award is modeled after the MBNQA, but includes awards for public-sector agencies and educational institutions as well as private-sector companies. It places an emphasis on human resource development and labor-management cooperation. The award is administered jointly by the New York Departments of Labor and Economic Development. There are two awards in each of three sectors: private (one-100 employees small; 101 or more employees large); public (one-500 employees small; 501 or more employees large); education (one-500 employees small; 501 or more employees large). Applications are available from the New York State Department of Economic Development and the New York State Department of Labor. The first awards were presented in May 1992 at a dinner hosted by governor Cuomo and lieutenant governor Lundine.

North Carolina

Contact

Mette Leather
Acting Director
Recognition Programs
North Carolina Quality Leadership Award
4904 Professional Court, Suite 100
Raleigh, NC 27609
T – 919-872-8198
F – 919-872-8199

Status

8/93 (Information date): The North Carolina Quality Leadership Award is modeled after the MBNQA and uses the previous year's Baldrige Award criteria and training materials. An executive order established the North Carolina Quality Leadership Awards Council, the body that presents the annual awards. The North Carolina Quality Leadership Award Foundation, a 501c(3) organization, administers the evaluation process. There are six categories: large manufacturing, large service, medium manufacturing, medium service, small manufacturing, and small service. A small firm must meet two of the three following criteria: not more than 100 full-time employees, not more than $5 million in sales, or not more than $7.5 million in total assets. It must also be able to document that it functions in the state independently of any equity owners. An honor roll also recognizes applicants that have demonstrated continuous quality improvement through implementation and deployment of prevention-based management and quality systems.

Oklahoma

Contact

Michael Strong
Executive Director
Oklahoma State Quality Award Foundation
6601 N. Broadway, Suite 244
Oklahoma City, OK 73116
T – 405-841-5295 or 800-879-6552
F – 405-841-5205

Status

4/04/94 (Information date): The Oklahoma Quality Award (OQA) was established in 1993 with the first awards scheduled for 1994. The award is patterned after the Baldrige Award. The award has six categories: large manufacturing, medium manufacturing, small manufacturing, large service, medium service, and small service. Any company operating in Oklahoma for three or more years may apply for the award. Some restrictions apply to subsidiaries and parent organizations. The 1994 OQA is available only for manufacturing and service organizations. Both for-profit and not-for-profit organizations can compete for the award. It is anticipated that the OQA will develop appropriate criteria by 1996 to allow awards in the governmental and educational sectors. An unlimited number of awards may be given in each category each year. Winners must be able to verify the quality practices associated with the applicant's major business in Oklahoma. Each applicant will receive a feedback report.

Oregon

Contact

Timothy Dedlow
Program Manager
Oregon Quality Award
7528 S.E. 29th Avenue
Portland, OR 97202-8827
T – 503-777-6057
F – 503-227-1599

Status

3/09/94 (Information date): The legislature passed a bill in July 1991 for an awards program recognizing firms with high-performance manufacturing practices. The Oregon Quality Initiative now has 501(c)3 status and has developed a business plan. The award is patterned after the Baldrige Award, but modified. The Oregon Quality Award requires applicants to complete a self-assessment to identify strengths and areas needing improvement. The award application materials include the Quality Self-Assessment Guide. First awards are to be given in October 1994. Award categories for 1994 will be private sector manufacturing and service. In 1995 education and government will be eligible as well. There will be no limit to the number of awards per year. Applicants will be judged against an exemplary level of quality standards. All applicants meeting or exceeding these standards will qualify for an Oregon Quality Award. The organization judged best of all the applicants that qualify for the Oregon Quality Award will then be evaluated against a superlative level of quality performance standards. If the standards are met, the organization will receive the

Governor's Award for Quality. Organizations of all sizes are eligible for the Oregon Quality Award. More than 98 percent of all Oregon businesses are made up of companies with fewer than 100 employees, and a special invitation is extended to those businesses to become involved in the award program.

Pennsylvania

Contact

Beverly M. Centini
Award Administrator
Pennsylvania Quality Leadership Foundation
P. O. Box 4129
Harrisburg, PA 17111-0129
T – 717-561-7100
F – 717-561-7104

Status

3/26/93 (Information date): On November 24, 1992, governor Casey signed into law a bill which established the Pennsylvania Quality Leadership Awards. Categories included in the program are large manufacturing, small manufacturing, large service, and small service organizations. Any publicly or privately held organization of any size, located in Pennsylvania, may apply. This includes, but is not limited to, state, local, and national government agencies, academic institutions, hospitals, healthcare units, financial institutions, service and manufacturing, religious, and nonprofit organizations. Subsidiaries, business units, divisions, or like components of large companies may be eligible for the award but must have existed in Pennsylvania for one year prior to the award application and must have a clear definition (for example, fiduciary responsibility) as reflected in corporate literature. The award is patterned after the MBNQA, but does not include the areas to address. The first awards are scheduled for October 25, 1994.

Rhode Island

Contact

Lynne Couture
Executive Director
Rhode Island Area Coalition for Excellence
P.O. Box 6766
Providence, RI 02940
or: 18 Emperial Place
Providence, RI 02903
T – 401-454-3030
F – 401-751-6703

Status

5/15/94: (Information date): Awards are now in place. Applications were due April 1, 1994. The criteria are patterned after the Baldrige Award, but modified. They are modeled to some degree after the Delaware criteria. The award categories are public (state and local government), private, and education. In each category there is a large and small (100 employees or less) division. There can be a maximum of two awards per category or a total of 12 awards to be given, although not all awards must be given. For nonwinners of the award, there can also be a Quality Achievement Recognition (certificate or plaque – format yet to be determined) given by the Quality Awards Council, which is under the Rhode Island Area Coalition for Excellence (RACE). There will also be on-the-spot Quality of Service Awards for individuals. These individual winners will be invited to a luncheon with the governor along with his or her employer.

Tennessee

Contact

Marie B. Williams
Director
Tennessee Quality Award Office
2233 Highway 75, Suite 1
Blountville, TN 37617-5840
T – 800-453-6474 or 615-279-0037
F – 615-279-0978

Status

3/94 (Information date): The Tennessee Quality Award is sponsored by the Tennessee Department of Economic and Community Development with administration assigned to the National Center for Quality Board with offices in Blountville and Washville. The award is modeled after the Baldrige Award. Any public or privately held organization of any size located in the state of Tennessee may apply. There will be four levels of application: Level 1 – Quality Interest: Recipients will have completed a series of educational steps which will provide an indicator of the seriousness of their interest in adopting and promoting quality; Level 2 – Quality Commitment: The intermediate level for organizations that have advanced from the knowledge and skills gained from their initial steps toward total quality improvement and progresses to a point of serious commitment; Level 3 – Quality Achievement: An advanced level of participation for organizations that have demonstrated, through their commitment and practice of quality principles, significant progress in building sound and notable processes deserving recognition; Level 4 – Governor's Quality Award: The highest level of recognition presented to organizations which have demonstrated, through their practices, the highest level of quality excellence. Applications were due on June 1, 1993. The first awards are planned for October 1993.

Texas

Contact

Robert Norwood
Executive Director
Quality Texas
17312 Whippoorwill Trail
Lago Vista, TX 78645
T – 512-267-2134

Status

11/17/93 (Information date): The Texas Quality Award is open to all businesses, state and local government agencies, nonprofit organizations, and educational institutions located in Texas. The categories (small – less than 100 employees, and large – more than 100 employees) have been developed for the following types of organizations: private: industrial organizations and service organizations; public: state and local government agencies/nonprofit organizations; and education: higher and K-12. The award is patterned after the MBNQA. The first awards will be presented at the award ceremony in April 1994.

Virginia

Contact

Elizabeth Holmes
Program Manager
U.S. Senate Productivity & Quality Award for Virginia
c/o VPQC/ISE
560 Whittemore Hall
Blacksburg, VA 24061-0118
T – 703-231-6100
F – 703-231-6925

Status

11/92 (Information date): Virginia does not have a separate state quality award, but uses the Senate Productivity Award with criteria that address quality issues. There are four categories: private sector manufacturing, private sector service, public sector state and federal agencies, and public sector local agencies. Applicants are evaluated based on: maturity of effort; top management commitment and involvement (leadership); employee involvement, development, and management of participation; recognition/rewards systems; plan for continuous improvement; performance measurement process (use of information); customer/supplier involvement; and results over time. An Award for Continuing Excellence is in place for previous medallion recipients. A previous winner must wait three years from receipt of last award before it is eligible to reapply. Applications are reviewed by an 18-member Senate Productivity Board. Applications are due each November and awards are presented the following April.

Foreign Quality Awards in Place

Argentina

Contact

Oscar A. Imbellone
Vice President
Fundacion Empresaria para la Calidad y la Excelencia
Santa Fe 846
2 Piso
1059 Buenos Aires
ARGENTINA
T – 311-8933/38 or 311-8941/46
F – 54-1-312-3228

Status

9/11/92 (Information date): The Premio Nacionel a la Calidad was signed into law in August 1992. The first award will be within 18 months of that date and will be presented by the president. A foundation is to be established and there will be a Board of Examiners. Categories have not yet been established. There will be up to two awards in each category.

Australia

Contact

Graham Spong
General Manager
Australian Quality Awards Foundation
P.O. Box 298
St. Leonards, N.S.W. 2065
AUSTRALIA
T – 011-61-2-439-8200
F – 011-61-2-906-3847

Status

1993 (Information date): The Australian Quality Awards Foundation has been formed to arrange the funding and expansion of the quality awards program. It is chaired by Allan Moyes, with a permanent secretariat run by Graham Spong. The New Australian Quality Prize was introduced in 1992 to recognize companies which achieve internationally competitive performance levels. To emphasize the importance of sustaining quality programs, the prize will only be awarded to winners of previous Australian Quality Awards competitions. There are no restrictions on the number of winners of the prize and the award.

Austria

Contact

Austrian Association for the Promotion of Quality
Gonzagagasse 2
A-Vienna 1010
AUSTRIA

Status

3/05/93 (Information date): The Ministry for Economic Affairs issues the National Quality Award for at least one, at the most up to three, products every year. The screening is done in several steps by means of a quality assessment system: (1) Companies which are convinced that one or several of their products are of exceedingly high quality apply for the Austrian Quality Seal (not yet "award"); (2) Quality standards for the various fields are determined by expert groups; (3) Official governmentally authorized testing institutes carry out the testing according to the previous quality standards. If the product passes the high quality level, the testing institute passes the high quality level, the testing institute certifies approval; (4) The Austrian Association for the Promotion of Quality then grants the Austrian Quality Seal to those products. Approximately 1000 companies are entitled to apply this seal; (5) Out of those 1000 companies a small number then applies for the National Quality Award which is very hard to achieve and which is finally determined by a jury; (6) The Ministry for Economic Affairs finally issues the National Quality Award as the highest product prize. The administration on behalf of the ministry is carried out by the Austrian Association for the Promotion of Quality.

Bahamas

Contact

Agatha K. Marcelle
Executive Director
Bahamas Quality Control
P.O. Box N-665
Nassau
BAHAMAS
T – 809-328-4310
F – 809-328-4131

Status

12/1/93 (Information date): The Bahamas has a "Staircase to Quality." The 100-Day Challenge acts as the first step in a three-tiered ladder. Companies are asked to do at least one thing better in the 100 days and are evaluated on the basis of how much they improved in the time frame for specific quality categories. Companies are evaluated at the beginning and the end of this 100-day period. The National Quality Award forms the next step in the quality process. In 1993 companies that had qualified for awards consideration in the 100-Day Challenge can participate in the National Quality Award. Here the emphasis will be put on the total quality level of the organization. They key difference is that, unlike the 100-Day Challenge where companies are evaluated in terms of the progress they have made in 100 days, companies applying for the National Quality Award are judged on the total quality offered to their customers and on the effectiveness of a five-year quality plan. Companies participating in the National Quality Award will be expected to have attained a high level of professionalism in survey research, statistical analysis, employee involvement, and operations planning. The Quality 365 Club is the final step on the staircase. Membership will be awarded on the basis of achievement of total quality service and products and will result after many years of competing in the National Quality Award. The criteria for this club is coverage and consistency. Coverage will determine inclusion in the club, while consistency will ensure that companies stay in the club.

Belgium

Contact

A. Giernaert
Vlaams Centrum voor Kwaliteitszorg
Secratariat Jury Commissie Quality Award
Researchpark Zellik
De Haak 2
B-1731 Zellik
BELGIUM

For the Walloon Region:
AWQ
Rue Puissant 15
B-6000 Charleroi

Status

5/26/93 (Information date): Flemish region: One of the three regions (Flanders) in Belgium has organized an award to promote quality, based on the criteria of the Baldrige Award.

3/18/93 (Information date): While Belgium does not have a national quality award, it is promoting a system of quality certification in Belgium. On a regional basis a quality award does exist.

Brazil

Contact

Basilio Dagnino
Technical Manager
Brazil National Quality Award Foundation
Av. Prestes
Haia, 733
s/1914
BR 01031
S. Paulo – SP
BRAZIL
T – 011-55-11-227-5216
F – 011-55-11-227-5216 or 991-2722

Status

2/92 (Information date): The criteria and schedule for Brazil's award have been patterned after the Baldrige Award. The first awards will recognize industrial, service, and small companies with outstanding quality performance, and will be presented during a ceremony in November by the president of Brazil.

Britain

Contact

Roy Knowles
British Quality Association
10 Grosvenor Gardens
London SW1W 0DQ
ENGLAND
T – 01-979-6511

Status

An award, created in 1984, is given annually to individuals and groups who have significantly improved the standard of quality of a product, process, or service. Any group or individual in the United Kingdom may be nominated for the award. Membership of the BQA is not necessary.

Canada

Contact

Susan Scissons
Total Quality Category Office
Canada Awards for Business Excellence
Service to Business Branch
Industry, Science & Technology Canada
235 Queen Street
Ottawa, Ontario K1A 0R5
CANADA
T – 613-954-4085
F – 613-954-4074

Status

4/6/93 (Information date): Canada has a quality category in the Canada Award for Business Excellence. The competition is open to profit-making business enterprises or their business units located in Canada in all fields of economic activity (natural resources, manufacturing, and services). It is patterned after the Baldrige Award. The award is given in recognition of outstanding achievement in overall business quality through a commitment to continuous quality improvement. The quality improvement must have taken place within the past five years, with the achieved level of quality performance being sustained over time. A random selection of external suppliers and customers of companies visited on-site may be contacted. There can be a maximum of three trophy winners. All trophy winners are accorded equal standing and are not ranked. A number of certificates of merit may also be awarded. The four categories of the 1993 evaluation criteria are: quality improvement policy and plan, implementation of policy and plan, results achieved, and future planning.

Canada – Quebec

Contact

Jean Fauteux
Association Quebecoise de la Qualite
Place Mercantile
770, Rue Sherbrooke Quest
10 Etage
Montreal (Quebec)
X3A 1G1
CANADA
T – 514-982-3008
F – 514-873-9912

Status

7/20/92 (Information date): The award is in very early planning stages.

Chile

Contact

Osvaldo N. Ferreiro
Coordinator
Productivity & Quality Program
Catholic University of Chile/Extension Center
Avda. Bernardo O'Higgens 390
Santiago
CHILE
T – 562-222-4516
F – 562-222-5916 or 562-222-5780

Status

10/31/91 (Information date): On October 28, 1991, the government of Chile (through the Minister of Economy) and the National Confederation of Production and Commerce (the association of all medium and big companies in the country) announced a joint effort to create, in 1992, a National Quality Award in Chile. Its levels include: (1) a technical council formed by representatives of the organizations related to quality, either public or private, to provide ideas and do some tasks; (2) a small working group with representatives from the government and the confederation.

Columbia

Contact

Teresita Beltran Ospina
Head of Standardization
Metrology & Quality Division
Ministerio de Desarrollo Economico
Calle 26 No. 13-19, Piso 34
Bogota
COLUMBIA
T – 571-334-0542
F – 571-281-1103

Status

7/17/91 (Information date): Columbia established a National Quality Award to recognize Colombian companies that excel in the quality achievement in quality management. At present, the main criteria for the award are ISO 9004 Part 1 – *Quality Management and Quality Systems Elements – Guidelines* and ISO 9004 Part 2 – *Guidelines for Services.*

Costa Rica

Contact

Office of Standards and Measuring Units
Ministry of Economy, Industry and Commerce
San Jose
COSTA RICA
T – 22 10 77

Status

2/15/93 (Information date): Costa Rica has not yet established a national quality award. However in 1991, it was given a Mention of Quality for same enterprises by the National Commission for Improvement of Quality. This commission is ascribed to the Min. of Science and Technology and created by Decree No. 20097, those enterprises were selected by organizations that deal with industry by their successes for improvement of the quality of their production. The national quality award will be included in the Law of Metrology, Standardization, Testing and Assurance of Quality that it will be approved soon. Costa Rica has said the Baldrige Award criteria will help Costa Rica to elaborate its regulations for its national quality award.

Czechoslovakia

Contact

V. Votapek
President
Czech Society for Quality
Novotneho Lavka 5
16668 PRAHA 1
CZECHOSLOVAKIA
T – 226845-9

Status

2/10/93 (Information date): A national quality award program in Czechoslovakia has been investigated during the last year under the responsibility of Federal Ministry of Control and CSQ. The program had not been approved by government due to the preparation of seclusion of Czechoslovakia. At present time, CSQ is fully responsible for this domain in the Czech Republic.

Denmark

Contact

Nils Agerhus
Head of Section
Office of Development Programmes
National Agency of Industry and Trade
Ministry of Industry
Tagensvej 137
DK-2200 Copenhagen N
DENMARK
T – 45-31-85-70-66
F – 45-31-81-70-68

Status

1/22/93 (Information date): No national quality award has been established at this time, but officials have been discussing and preparing some kind of quality or productivity award for some time in relation to some of their programs. In the spring of 1993 they are planning a program titled "the Knowledge and Quality Program for Small and Medium-Sized Companies." It is their plan that there should be some kind of award in relation to this program. The program will start around July or August 1993.

European

Contact

Mike Gallagher
Manager
European Quality Award
The European Foundation for Quality Management
The European Quality Award Secretariat
Avenue des Pleiades 19
1200 Brussels
BELGIUM
T – 32-2-775-3511
F – 32-2-779-1237

Status

4/6/93 (Information date): First awards were given in October 1992. Applications for the European Quality Award will be assessed on the following criteria: leadership, policy and strategy, people management, resources, processes, customer satisfaction, people satisfaction, impact on society, and business results. In general, the award scheme will be for companies and subsidiaries or divisions of companies, based in Western Europe. Applicants are encouraged to submit an eligibility form as early as possible in the year they wish to submit their application. The European Quality Award for 1993 incorporates: (1) European Quality Prizes, awarded to a number of companies that demonstrate excellence in the management of quality and their fundamental process for continuous improvement, and (2) The European Quality Award, which is awarded to the most successful exponent of TQM in Western Europe (the best of the prize winners).

Finland

Contact

The Finnish Society for Quality

Status

2/26/93 (Information date): (Information provided by the Ministry of Trade and Industry.) The Finnish Society for Quality grants quality awards. The rules for the awards, more or less reflecting the Baldrige Award and similar European awards, were adopted in 1990. Represented on the board annually deciding the Award(s) to be given are the Ministry of Trade and Industry, the Federation of Finnish Industries, the Central Chamber of Commerce, the Finnish Standards Association, and the Finnish Society for Quality. Three company categories are available for choice: industrial enterprises, service enterprises, and SMEs can, if they so wish, participate in one of the two general categories. Eligible candidates are any public or private industrial, administrative, or service corporations aiming at profitable operation, which have been operating in Finland for at least five years. In a company with more than 300 persons on its payroll, any of its independent branches may participate on its own account. Willingness to participate is announced through an application lodged with the Society for Quality, from which detailed information is also available. A firm winning an award in one year cannot participate the next year. A separate visit is made to a few firms among which the board expects the winner(s) to be found in the light of the applications. The evaluation is carried out by a team of experts. All participants receive a written statement on the quality aspects as involved in their operations. All corporate details are dealt with confidentially. The board sees that none of the applications is evaluated by any of the firm's competitors. The criteria are measures of the management, information and its analysis, strategic quality planning, personnel development and participation, QC relative to operative processes, qualitative and operative results, client-orientedness and clients' satisfaction, and favorable effects for the society.

France

Contact

Miss De Tastes
French Association for Quality
Association Francaise pour la Qualité
Tour Europe, Cedex 7
92080 Paris La Defense
FRANCE
T – 011-33-1-42915953
TELEX 611974 f

Status

1989 (Information date): An award is in place, but no award was given in last two years.

India

Contact

Director General
Bureau of Indian Standards
9, Bahadur Shah Zafar Marg
New Delhi 11002
INDIA

Status

5/09/94 (Information date): The Rajiv Gandhi National Quality Award was instituted by the Bureau of Indian Standards in 1991. This award has been designated in line with the Baldrige Award, Deming Prize, and European Quality Award. There are four awards consisting of one for large-scale manufacturing units, one for small-scale manufacturing units, one for service sector organizations, and one for best of all. In addition, there are six commendation certificates each for large-scale and small-scale manufacturing units as per the following industrial sectors: (1) metallurgical industry, (2) electrical and electronic industry, (3) chemical industry, (4) food and drug industry, (5) textile industry, and (6) engineering industry and others. All manufacturing units and service sector organizations located in India are eligible to compete for the awards and commendation certificates. They will have to submit a copy of certificate of incorporation or similar other document(s) authenticating the name of the unit/organization and its premises. Definitions of large-scale and small-scale industry are the same as laid down by the government of India. For the purpose of this award, the manufacturing unit shall be: "A manufacturing unit situated at one place which can be a complete firm or a unit (division of a firm housed at one location)." The criteria categories are leadership, policies and strategies, human resource management, resources, processes, customer satisfaction, employee satisfaction, impact on society, and business results.

Contact

Najma Heptulla
National Chairman
Institute of Directors
"Machuban Building"
107/55 Nehru Place
New Delhi 100 019
INDIA
T – 91-11-6421668
F – 91-11-6468537

Status

5/20/92 (Information date): The Institute of Directors (IOD) is a registered non-for-profit apex body of directors, committed to improving the competitiveness of Indian business and industry through total quality strategies. The IOD has instituted a National Quality Award for outstanding achievement in the pursuit of total quality in the Indian industry on the pattern of the Baldrige Award. The first winners were announced in 1991.

International

Contact (in the United States)

Ed Fuchs
International Academy for Quality
AT&T Bell Laboratories
Crawford Corner Road
Room #2M-524
Holmdel, NJ 07733
T – 201-949-6244

Contact

Roy Knowles
International Academy for Quality
IAW Admin. Office
c/o Deutsche Genselschaft for Quality, e.v.
August Schanz Strausse 21A
Frankfurt am Main 50
FDR
T – 11-44-81-979-6511

Status

6/1/90 (Information date): The awards, established in 1991, are for (1) Technology paper – for the best nominated quality technology paper published during the preceding two-year period; (2) Management paper – for the best nominated quality management paper published during the preceding two-year period; and (3) Individual achievement – for the best nominated individual quality achievement during the preceding two-year period. The IAQ has initiated its Quality Award Scheme to give international recognition for outstanding individual contributors to the promotion of a better understanding of the need for product and service quality improvement. Any individual, other than an academician of the IAQ, may be nominated for the award.

Israel

Contact

Avigdor Zonnenshain
Quality and Productivity Director
Association of Electronics Industries
P.O. Box 50026
Tel-Aviv 61500
ISRAEL

Status

12/18/91 (Information date): The Quality Prize in Israel has been established by the AEI. The president of the state of Israel has agreed to take the award ceremony under his auspices. The competition is open to corporations and enterprises of all sizes, and to subunits of corporations, provided these encompass at least one particular business area and constitute business units – they encompass a range of activities characteristic of a selfstanding business. The prize is intended for award around Hanukkah time – at the end of the fiscal year, with the first award scheduled for December 1991 or January 1992.

Japan

Contact

Kohei Suzue
President
Union of Japanese Scientists and Engineers
5-10-11 Sendagaya, Shibuya-ku
Tokyo 151
JAPAN
T – 03-352-2231
F – 03-356-1798

Status

The Deming Prize was instituted in December 1950 in honor of W. Edwards Deming.

Malaysia

Contact

Annie Thomas
Pusat Daya Pengeluaran Negara
National Productivity Centre
(Ministry of Trade & Industry)
Peti Surat 64, 46904 Petaling Jaya
MALAYSIA
T – 755-7266 (15 tallan)
F – 03-757-8068

Status

4/4/90 (Information date): Malaysia is in the process of formulating guiding principles for an award in Malaysia, similar to the Baldrige Award.

Mexico

Contact

Raul Macias G.
General Director
Mexico National Quality Award
Heriberto Frias #249
Col. del Valle
Del. Benito Juarez
C.P. 03100 Mexico, D.F.
P.O. Box: 12700
MEXICO
T – 91-5-639-67-48 or 91-5-639-68-63
F – 91-5-639-61-70

Status

2/18/92 (Information date): First awards were given in 1990. This award acknowledges the effort Mexican organizations make toward total quality under the following categories: large industrial companies, medium or small industrial companies, large commercial companies, medium or small commercial companies, large service companies, and medium or small service companies. Ten acknowledgments may be given out, with no more than two for each of the categories mentioned (some categories may not be considered available). The decree established that the president of Mexico will deliver this award to the winning organization each year.

Netherlands

Contact

Jan Balk
Nederlandse Kwaliteit
Postbus 205
2000 AE Haarlem
NETHERLANDS

Status

1/18/93 (Information date): The Dutch Quality Prize, a slightly modified version of the European Quality Award (separately for profit and nonprofit organizations), will be given in November 1993. Unlike the Baldrige Award, the Dutch Quality Prize is an initiative of a private organization (but with full government support): "Nederlandse Kwaliteit," a council of managers with experience in the introduction of quality management.

New Zealand

Contact

Doug Matheson
Chairperson
New Zealand Quality Awards Foundation
"Karinga"
West Bush Road
RD 8
Masterton
NEW ZEALAND
T – 00-64-6-377-4111
F – 00-64-6-377-4111

Status

8/27/92 (Information date): Following consultation with the business sector and examination of overseas models, it was determined that the New Zealand National Quality Award (NZNQA) should be a premier award with trans-Tasman and international standing, and should acknowledge both excellence in the management of quality, and provide participating firms with a benchmark against which they could measure their approach to quality. A privately funded quality foundation is being established to lead the quality awards and to generate funding and support. The foundation, while not yet established as a legal entity, comprises 40 companies. It is proposed that the first NZNQA be presented at the end of 1993. The Baldrige Award criteria will be used as the basis for the NZNQA.

Northern Ireland

Contact

Roy Adair
CED/Director
Northern Ireland Quality Centre
Midland Building
Whitla Street
Belfast BT15 1NH
IRELAND
T – 0232-352999
F – 0232-352888

Status

3/18/93 (Information date): The Northern Ireland Quality Awards are organized by the Northern Ireland Quality Centre, with sponsorship from Gallaher, the IDB, and LEDU. The awards are open to all organizations in the public, private, and voluntary sectors which have a base in the Province. In 1991 the award scheme aimed to attract a wider application following a restructuring of the categories into three sections. These are manufacturing, service, and small organizations (classed in this instance as having less than 50 employees). The winners will be entitled to send a representative on a fact-finding mission to internationally recognized quality companies. The criteria is the same as the Baldrige Award plus one category. The questionnaire asks questions in eight areas: leadership, use of data and measures, planning for quality improvement, people, quality assurance, continuous improvement, customer satisfaction, and quality results. The application is eight pages. There are members of the Board of Examiners (same each year). The site visit is one day. The awards are presented to the winners by Patrick Mayhew, secretary of state for Northern Ireland. Currently, the budget is £18,000.

Norway

Contact

Norwegian Society for Quality
Brynsveien 96
1352 Kolsas
NORWAY

Status

2/16/93 (Information date): Via the Royal Ministry of Industry and Energy, the Norwegian Society for Quality administers a quality award.

Philippines

Contact

Philippine Society for Quality Control
Suite 607
Campos Rueda Building
Urban Street
Makati
Metro Manila
PHILIPPINES

Status

6/1/90 (Information date): A joint project with the Philippine Society for Quality Control and National Quality Campaign is in place. The objective of the Outstanding Quality Company of the Year Award is to stimulate interest in the advancement and dissemination of quality management practices in the industry, service, and other areas by giving due recognition to the company which has shown outstanding achievements through exemplary quality improvements. The nominated company must be a registered business enterprise under laws of the Philippines.

Singapore

Contact

Lee Siew Hon
The Singapore Quality Award Info. Centre
c/o National Productivity Board
NPS Building
2 Bukit Merah Central
SINGAPORE
0315
T – 279-3708 or 279-3711
F – 279-6667

Status

2/14/94 (Information date): The Singapore Quality Award seeks to recognize organizations for outstanding quality achievements. Jointly initiated by the National Productivity Board and the Singapore Institute of Standards and Industrial Research, the award has the patronage of the prime minister and the support of the private sector. All publicly and privately registered organizations are eligible to apply. However, private organizations applying for the award must have a major business operation located in Singapore which can be assessed by a panel of quality professionals. Subsidiary companies applying for the award are governed by specific conditions. Each award winner will receive a specially crafted award trophy at a grand presentation ceremony. Part of the objective of the award is to encourage companies to develop a quality focus, therefore all winners will share their quality success with others through seminars and workshops. Details of the award criteria will be released in 1995. The criteria include customer focus and satisfaction, leadership and quality culture, human resource development and management, strategic quality planning, management of process quality, information and analysis, and quality and operational results.

Spain

Contact

Spanish Association for Quality
Zurbano, 92
1 Drcha.
28003 Madrid
SPAIN
T – 91-441-75-43
F – 91-441-73-90

Status

2/2/93 (Information date): For the National Quality Award, the Ministry of Industry, Commercialization & Tourism has delegated the functions of the secretary to the Spanish Association for Quality. There are two awards only: companies or organizations with more than 250 employees or small businesses or organizations with 250 employees or less. Neither may be awarded if no one qualifies. These are 18 areas to address. Companies must submit 17 of the 18. The Evaluation Committee reserves the right to verify the application contents through whatever systems it deems necessary – site visits, requests for more information, verbal explanations, third parties, and so on. The nominated applications go on the board of judges. Winners share information with other Spanish organizations to benefit national competitiveness.

Sweden

Contact

Jonny Lindstrom
President
Swedish Institute for Quality
Gardatorget
S-412 50 Goteborg
SWEDEN
T – 011-46-31-773-0875
F – 011-46-31-773-0645

Svante Lundin
Counselor, Science & Technology
Sveriges Tekniska Attacheer
(The Swedish Technical Attache System)
Swedish Embassy
600 New Hampshire Ave., N.W.
Washington, DC 20037
T – 202-337-5186
F – 202-337-6108

Status

11/28/92 (Information date): The Swedish Institute for Quality has been established to manage the Swedish Quality Award and other quality programs. The Swedish government and Swedish industry and public services sponsor the institute. The first awards were delivered in December 1992 by king Carl XVI Gustaf of Sweden. A comprehensive feedback report presents the evaluation results to each applicant company.

Taiwan

Contact

Kao Sin-Yan
Chairman
Center Satellite Development (CSD)
National Quality Award Section
CSD Industrial Coordination Center
7th Floor, No. 5, Tun-Hwa North Road
Taipei
TAIWAN
T – 886-2-751-3468
F – 886-2-781-7790

Status

1/29/93 (Information date): The National Quality Award was established by the Ministry of Economic Affairs (MOEA) in 1990 and was formally authorized by its Executive Yuan. The primary purpose of the award is to award outstanding achievement in the area of TQM, as well as to further the upgrading of quality levels. The Working Group of the National Quality Award Examining Committee CSD administers the award. It is managed by the Industry Development Bureau, Ministry of Economic Affairs, and The National Quality Award Examining Committee. The application period is January-February of each year. There is no fee charged. Categories and eligibility are enterprises – all licensed manufacturers, small and medium enterprises – all licensed manufacturers, in conformance with MOEA's qualified standard for small and medium enterprise; and individuals – Taiwan's citizens. A maximum of two awards per category may be given each year. The Criteria are: Enterprises: (1) management concepts, goals, and policies, (2) organization and operation, (3) human development and utilization, (4) information management and utilization, (5) research and development, (6) quality assurance, (7) quality of customer service, (8) public responsibility, (9) TQM performance; Small and medium enterprises: (1) management concepts, goals, and policies, (2) organizations, operations, and human development, (3) research and development, information management, (4) quality assurance, (5) public responsibility, (6) TQM performance; Individuals: For the results achieved due to researching or promoting TQM, contributions to the country and society in the same area. The award ceremony is held each September and winners will receive a National Quality Award certificate and a bronze trophy.

United Kingdom

Contact

J. N. Slaughter
Secretary
Quality Award Committee
61 Southwark Street
London SE1 1SA
UNITED KINGDOM
T – 071-928-8999
F – 071-928-5566

Status

8/7/92 (Information date): The Quality Award Committee has completed its work and has prepared a final report to the president of the Board of Trade. The committee recommends that a new high profile U.K. total quality award be established and awarded for the first time in 1994. The committee recommends the new award should use the European Quality Award's criteria. The proposed name for the award is The Prime Minister's Award for Total Quality.

Uruguay

Contact

Ruperto E. Long
Exec. Coordinator
Comite Nacional de Calidad
Galicia 1133 Piso 1
Montevideo
URUGUAY
T – 98 08 66
F – 98 44 32

Status

6/25/92 (Information date): The president of Republica Oriental del Uruguay, Luis Alberto Lacalle, launched in May a National Quality Committee (CNC). The CNC is in charge of a National Quality Program. It is composed of three subprograms: education in quality, National Quality Award, and users revolution. The main idea of the National Quality Award is to reward those enterprises which are making outstanding efforts to adopt modern techniques of quality management. Uruguay is considering adapting the ideas behind the Baldrige Award.

Appendix C:

The International Academy for Quality

Constitution (Issue April 1990)

Article 1. Title, Headquarters, Fiscal Year

The institution is known as **International Academy for Quality,** for short: The IAQ.

The members of the IAQ are known as **Academicians.**

IAQ is a non-profit organization incorporated. The **Headquarters** is located in Frankfurt a. Main, Germany, as a Registered Association IAQ e.V.

The **fiscal year** is the calendar-year.

Article 2. Aims, Principles, Objectives and Strategies

2.1 The **aims** of the IAQ are to promote research into the philosophy, theory and practice of all activities involved in achieving quality for the best use of world resources, and to develop a spirit of mutual comprehension and cooperation at both the national and international levels.

2.2 The **Principles** of IAQ are:

a) IAQ is organized on a strictly international basis, the Academicians represent the best quality knowledge and practices in all regions of the world.

b) IAQ membership consists of individuals who have been recognized as having achieved high standards in the quality field and who work actively in terms of the criteria established for Academicians.

c) IAQ's effect, visibility and influence is achieved through the individual work and visibility of the Academicians themselves.

d) IAQ's activities are carried on in a thorough, clear, organized work process.

2.3 The **objectives** of the IAQ are:

a) to detect, bring forward, encourage and facilitate every valuable and promising activity initiated by national or international organizations, or by committees, or by individuals which is likely to further the foregoing aims.

b) to provide organizations, specialists or not in quality, with a consultative body competent to advise on quality matters without interfering with their internal affair.

In order to reach these objectives the IAQ will operate by giving guidance to other organizations rather than by acting directly itself.

2.4 The **strategies** to reach these objectives include:

a) Sponsoring and providing themes and selecting papers for congresses or conferences organized by national or international organizations dealing with quality,

b) facilitating the cooperation between technical committees of national and international organizations,

c) publishing or encouraging the publication and distribution of significant quality-related material, e. g. in an IAQ-Book Series,

d) defining, implementing and/or sponsoring quality-related projects,

e) Annual meetings of Academicians participating in international conferences, aimed at restating the most recent developments and research in quality on a worldwide basis, and

f) triennial Assemblies of academicians (see Art. 5)

Article 3. Membership

3.1 Election of Academicians
To achieve the objectives in a continuously changing world with the rapidly evolving techniques of management, requires a small group comprising the most active and experienced protagonists of quality in the world.

The academicians are selected from amongst the best experts operating in the three major regions of the world:

A) America-North, Central, South
B) Europe, Africa
C) Asia, Australia, Pacific

Most of them will be practitioners having a direct personal experience in managing the quality activities of an enterprise. They will be supplemented by specialists in quality techniques and by experts in management. In order to facilitate the selection of candidates, their ability and their willingness to contribute actively to the visibility of the Academy are the primary considerations.

Admission as an Academician shall be regarded as a high honor. The election procedure is determined by the Rules of Procedure.

3.2 Active Members

Active membership implies the obligation to participate actively and visibly to the publicity of the whole world, in promoting the objectives and strategies of the Academy and a willingness to voluntarily devote an adequate time of IAQ-activities.

3.3 Corresponding Members

After years of active membership it is recognized that circumstances may no longer allow an active member to travel abroad and participate in meetings, but whose opinions and contributions in writing remain valuable to the objectives and the visibility of the Academy. Members in such a situation may be permitted by agreement to the Nominating Committee the status of "Corresponding Membership." These members have no voting rights.

3.4 Emeriti and Honorary Members

As a recognition of distinguished service rendered to the Academy, Academicians can be promoted to Emeritus-Status – Honorary Academicians can be selected, whose experience and prestige would make them valuable to the IAQ.

No specific activity is required from Emeriti or Honorary Academicians, nor do they have voting rights.

3.5 Termination of membership

When circumstances preclude an academician from active participation in IAQ-affairs, that Academician may terminate membership by letter to the President. Termination may also take place in accordance with special rules of membership status, outlined in the Rules of Procedure.

Article 4. Authority and Organs

The authority for governing the IAQ is vested in the whole Academy. Organs of IAQ are:

- The Assembly of Members (see Art. 5)
- The Board of Directors (see Art. 6)
- The Officers Committee (see Art. 6)

Article 5. The Assembly of Members (Triennial)

a) The Assembly of Members shall convene regularly, at least once in every three years. The Assembly can be convened additionally by a move (in writing) of two officers or by the request of a minimum one quarter of the Academicians.

b) Academicians will be invited to attend an Assembly of Members and the agenda will be announced in writing by the President, at least 30 days in advance. The Assembly is empowered to pass resolutions, if one third of the total membership are present or represented.

c) The Assembly of Members will be chaired by the Chairman of the Board of Directors or an Academician appointed by him/her. The Assembly decides on the following affairs:

- Reports and Discharge of the Executives

- Election of Board members and Officers

- Fixing of Membership Fees

- Additions and Amendments of Constitution (see Art. 8)

- Dissolution (see Art. 9)

d) For voting in the assembly each active member can, by submitting written authorization, be represented by another active Academician. Each Academician can represent only one other.

e) All resolutions will be recorded in the Minutes and be signed by the President and the secretary of the meeting.

Article 6. The Board of Directors, Officers, Program-Directors

The IAQ is manage by two committees and assisted by Program-Directors:

6.1 The Board of Directors

The Board of Directors consisting of the Chairman and no less than three Directors is in charge of

- Overall Policy of the Academy, including Policy Formation and Direction,

- Long Range Planning,

- Nomination and Election of Academicians, Growth Regulation and Control.
- Preparations for the Election of the IAQ-Officers, and
- Trustee Activities

6.2 The Executive Committee

The Executive committee consists of four officers: The President, the Vice President for Membership and education, the Vice-President for Technical Activities and the Vice-President for Publications, Conferences and Relationship. They are in charge of both policy implementation and leading of all activities of the Academy.

6.3 Election and Term of Office

a) It is the Academy's intent to always select the best candidate for each officer's position. An important prerequisite for election to office in IAQ is the willingness and ability of a candidate to devote a reasonable amount of time to such office. Every effort will be made to have a balanced representation on the Executive Committee and the Board of Directors from the three major regions of the world mentioned in Art. 3.1.

b) The term of office for newly elected Officers and Directors begins at the first of January, following the election, and lasts for three years.

6.4 Program-Directors

Every Academician may be assigned specific duties as Program Director in both administrative and technical fields, and active participation in the performance of these duties is a vital role for each Academician.

Article 7. Finance and Administration

7.1 Finance

a) The IAQ is based on goodwill and devotion to its aims from Officers and Academicians, who receive no remuneration and who themselves undertake the necessary minimal clerical services.

b) Active and corresponding Academicians pay annual fees fixed by the assembly of members. – Other sources of the Academy funds are donations, royalties, etc.

c) International Associations active in quality bring to the IAQ their moral support and their material cooperation for achieving the objectives mentioned in Article 2, and in particular for conferences.

7.2 Administration

a) The Administrative office in the IAQ-Headquarters in Frankfurt undertakes those administration activities described in detail in the Rules of Procedure.

b) The detailed functioning of the IAQ is guided by Rules of Procedure. These Rules of Procedure can be amended by a decision of the Executive Committee in accordance with the Board of Directors. A motion to amend can also be initiated by a group of no less than 10 active Academicians.

Article 8. Constitutional Change

A constitutional change will be implemented in the following manner:

a) The Board of Directors and the Executive Committee can introduce a motion for change by unanimous decision. A motion for change of the Constitution can also be initiated by no less than 20 active Academicians acting as a group.

b) The motion has to be submitted for ballot in writing to the active membership of the Academy. If after three months from the date of submission of the ballot to the membership at least 2/3 of the received ballots are in favor of the change it is deemed accepted and in effect.

Article 9. Dissolution

a) The Academy can be dissolved only by written ballot from the whole Academy. The procedure is the same as in Article 8, with the proviso that at least 2/3 of all active Academicians must respond and at least 2/3 of them must be in favor of dissolution. If less than 2/3 of the active Academicians respond, a second ballot is conducted not earlier than 6 months after the first. In this second ballot the 2/3 majority applies regardless of the number of answers received.

b) Should there be, for any reason, the dissolution of IAQ, any and all assets will be turned over, in accordance with suitable procedures, to a non-profit organization having similar aims and objectives.

History

Since 1966 there were ongoing considerations on improvement of international relations between ASQC, the European Organization for Quality Control (EOQC), and JUSE. That led to the foundation of the IAQ in 1969 by 21 founding members.

In 1975, the academy reached full operation. In 1990, 61 members of 23 countries on four continents belonged to it. The academy is managed by the Executive Committee which changes in three-year terms.

Members of the Executive Committee since 1975 are listed in Figure C.1. Members of the Executive Committee in 1991-1993 are listed in Figure C.2 and members of the Executive Committee in 1994-1996 are listed in Figure C.3. See Figure C.4 for a list of addresses. Figure C.5 provides a regional distribution of IAQ members as of January 1994.

	President	Vice President	Seceretary-General
1975	A. V. Feigenbaum	K. Ishikawa	U. Turello
1978	W. Masing	C. Bicking	Y. Kondo
1981	K. Ishikawa	E. Blanco	M. Liebman
1984	M. Liebman	Y. Kondo	H. Zeller
1987	H. Zeller	J. Harrington	M. Imaizumi

Figure C.1. Members of the Executive Committee since 1975.

President . . . . J. Harrington

Vice President for Membership and Education . . . . W. Golomski

Vice President for Technical Activities . . . . A. Aune

Vice President for Publications, Conferences and Relations . . . . M. Ito

Figure C.2. Members of the Executive Committee 1991-1993.

President . . . . Y. Kondo

Vice President for Membership and Education . . . . W. Golomski

Vice President for Technical Activities . . . . V. Seitschek

Vice President for Publications, Conferences and Relations . . . . Y. Bester

Figure C.3. Members of the Executive Committee 1994-1996.

President
Yoshio Kondo
29 Higashi-Takagicho
Shimogamo, Sakyo-ku
Kyoto 606
Japan

Vice Presidents
Yoseph Bester
President
Prime Institute
2 Yehuda Gur Street, Hod Hacarmel
Haifa 34987
Israel

William A. J. Golomski
President
W. A. Golomski Associates
20 East Jackson Boulevard, Suite 850
Chicago, Illinois 60604-2208

Victor Seitschek
Magdalenastrasse 20
4040 Linz
Austria

IAQ Administrative Office
Horst Fuhr
c/o Deutsche Gesellschaft fur Qualitat (DGQ)
Mail Box 50 07 63
W 6 Frankfurt a. Main 50
Germany

Figure C.4.   Addresses of Executive Committee members.

Argentina

Marcos E. J. Bertin, Buenos Aires
Enrique Jorge Garcia, Buenos Aires
Juan C. L. Musi, Buenos Aires

Australia

R. M. Burt, Eastwood NSW
Phillip A. Richardson, Melbourne

Austria

Victor Seitschek, Linz

Brazil

Gabor S. Aschner, Rio de Janeiro
Mauro Luis Correia, Sao Paul – SP
Masao Ito, Rio de Janeiro

Canada

Denys A. Pilon, LaSalle

Czech Republic

Agnes Zaludova, Prague

Denmark

Ove Hartz, Lyngby
Carl A. Kofoed, Nordborg

Finland

Juhani J. Koivula, Espoo

France

Alain Michel Chauvel, St. Cyr/Dourdan

Germany

Wolfgang Hansen, Krailling
Dietmar C. H. Mangelsdorf, Geretsried
Walter Masing, Erbach
Hermann Zeller, Groebenzell

India

K. Pal Basanta, Bangalore

Ireland

John A. Murphy, Dublin

Israel

Yoseph Bester, Haifa

Italy

Tito Conti, Ivrea
Giovanni U. Mattana, Milano

Japan

Yoji Akao, Tokyo
Masashi Asao, Osaka
Masumasa Imaizumi, Tokyo
Noriaki Kano, Tokyo
Yoshio Kondo, Kyoto
Hitoshi Kume, Tokyo
Ikuro Kusaba, Fujisawa
Shoichi Shimizu, Kyoto

Lichtenstein

H. D. Seghezzi, Schaan

Figure C.5. Regional distribution of IAQ members (as of January 1994).

Norway

Asbørn Aune, Trondheim-NTH

Phillippines

Miflora Gatchalian, Quezon City

Portugal

Antonio de Almeida Jr., Lisboa

Spain

Enrique Blanco Loizelier, Madrid

Sweden

Kerstin M. C. Joenson, Moelnlycke

Lennart Sandholm, Djursholm

Switzerland

Enrique Sierra, Geneva

United Kingdom

Norman T. Burgess, Egham

John A. Goldsmith, Hertford

John Groocock, Petts Wood

Roy Knowles, Hampton

United States

Kenneth E. Case, Stillwater, OK

Robert E. Cole, Berkeley, CA

Armand V. Feigenbaum,
 Pittsfield, MA

R. L. Fiaschetti, Whittier, CA

Edward Fuchs, Holmdel, NJ

A. Blanton Godfrey, Wilton, CT

William A. J. Golomski, Chicago, IL

Eugene L. Grant, Palo Alto, CA

James Harrington, San Jose, CA

John D. Hromi, Rochester, NY

Walter L. Hurd, Jr., Los Altos, CA

Spencer Hutchens, Jr.,
 Rolling Hills, CA

Murray E. Liebman, Sacramento, CA

David B. Luther, Corning, NY

Julius Y. McClure, Weatherford, TX

Richard J. Schlesinger,
 Los Angeles, CA

Dorian Shainin, Manchester, CT

Kenneth S. Stephens, Marietta, GA

Raymond Wachniak, Brookfield, WI

Figure C.5 *(continued).*

Index